McGraw-Hill

My Math

Interactive Guide

Grade 4

McGraw Hill Education

ConnectED.mcgraw-hill.com

Mc Graw Hill Education

STEM McGraw-Hill is committed to providing instructional materials in Science, Technology, Engineering, and Mathematics (STEM) that give all students a solid foundation, one that prepares them for college and careers in the 21st century.

Send all inquiries to:
McGraw-Hill Education
8787 Orion Place
Columbus, OH 43240

Selections from:
ISBN: 978-0-02-130755-5 *(Grade 4 Student Edition)*
MHID: 0-02-130755-5 *(Grade 4 Student Edition)*
ISBN: 978-0-02-131406-5 *(Grade 4 Teacher Edition)*
MHID: 0-02-131406-3 *(Grade 4 Teacher Edition)*

Printed in the United States of America.

4 5 6 7 8 9 ROV 20 19 18 17 16

Contents

Chapter 5 Multiply with Two-Digit Numbers

Chapter 6 Divide by a One-Digit Number

Chapter 7 Patterns and Sequences

Chapter 8 Fractions

Chapter 13 Perimeter and Area

Chapter 14 Geometry

NAME ______________________________ DATE ______________

Lesson 4 Note Taking

Order Numbers

Read the question. Write words you need help with and research each word. Use your lesson to write your Cornell notes. Write or draw math examples to explain your thinking. Share your examples with a classmate.

Building on the Essential Question	**Notes:**
How does a place value chart help order numbers?	Write the numbers being ordered with each ________ on a place-value chart. Start with the greatest ______ ______ position. Compare using the ___________ than symbol (>). Compare the ________ in the next ______ ______ position. Continue to compare until the ________ are different.
Words I need help with:	

My Math Examples:

NAME ______________________ DATE ____________

Lesson 5 Concept Web

Use Place Value to Round

Use the concept web to write examples of rounding the number 284,761 to different place values.

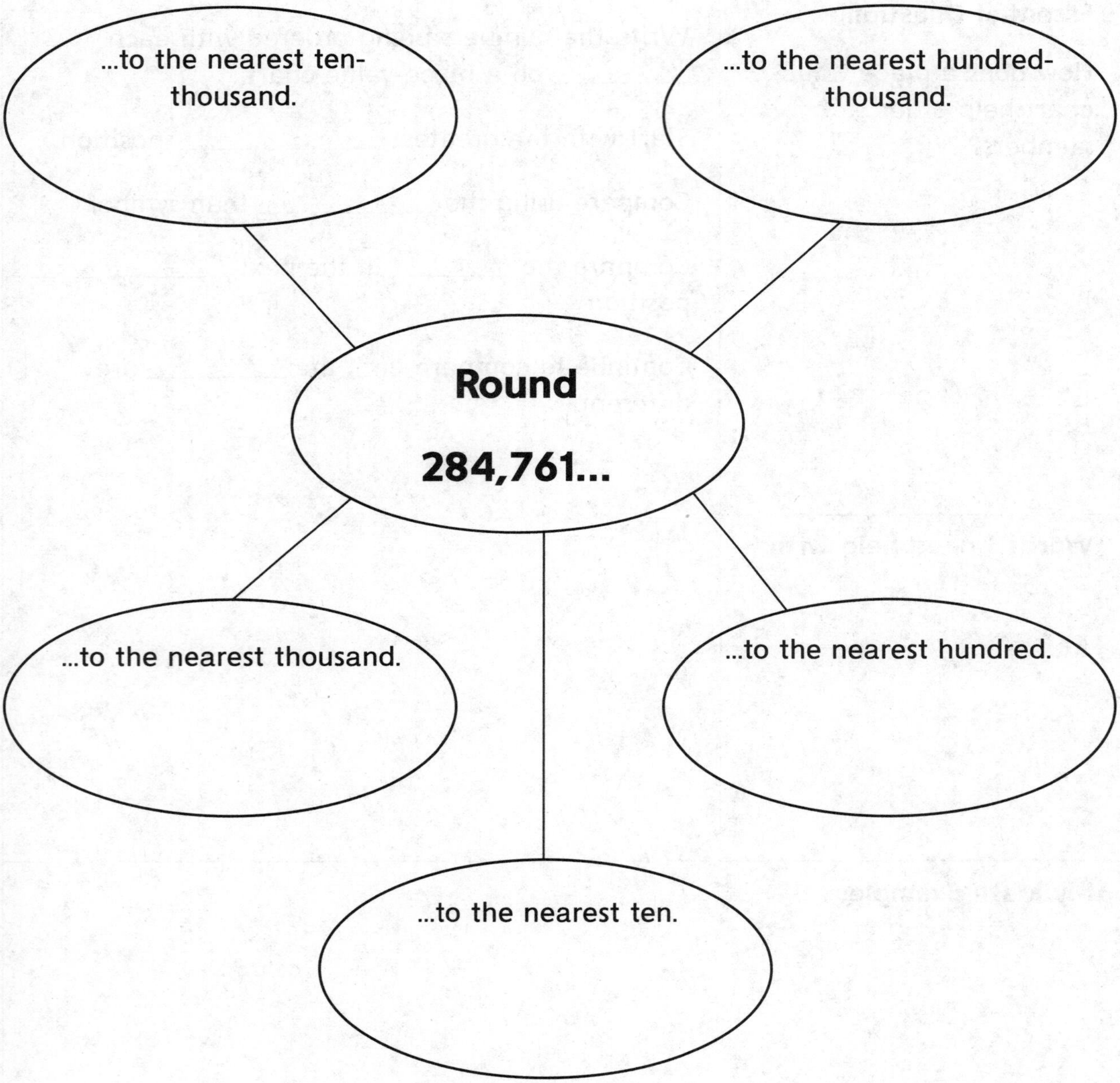

NAME ______________________ DATE ______________

Lesson 6 Problem-Solving Investigation

STRATEGY: Use the Four-Step Plan

Prices of Cars	
Cars	Price
Car A	$83,532
Car B	$24,375
Car C	$24,053
Car D	$73,295

Solve each problem using the four-step plan.

1. **Mr. Kramer** is buying a car.
The list of **prices** is shown in the table.
Mr. Kramer wants to **buy** the **least** **expensive** car.
Which car should **he** (Mr. Kramer) buy?

Understand

I know...

I need to find...

Solve

Thousands			Ones		
hundreds	tens	ones	hundreds	tens	ones

Plan

I can use a ______________________
to help me solve.

Check

2. **A restaurant** made **more** **than $80,000** but **less** **than $90,000** **last** **month.**
There is a **6** in the **ones** place, a **3** in the **thousands** place,
a **7** in the **hundreds** place, and a **1** in the **tens** place.
How much money did the restaurant make **last** **month?**

Understand

I know...

I need to find...

Solve

Thousands Period			Ones Period		
hundreds	tens	ones	hundreds	tens	ones

Plan

I can use a ______________________
to help me solve.

Check

NAME ______________________ DATE ____________

Chapter 2 Add and Subtract Whole Numbers

Inquiry of the Essential Question:

What strategies can I use to add or subtract?

Read the Essential Question. Describe your observations (I see...), inferences (I think...), and prior knowledge (I know...) of each math example. Write additional questions you have below. Then share your ideas and questions with a classmate.

1,255	rounds to →	1,000
+ 6,740	rounds to →	+ 7,000
		8,000

I see ...

I think...

I know...

---

thousands hundreds tens ones

```
      8 13
  5, 9̸ 3̸ 8
–     2 7 6
  5, 6 6 2
```

I see ...

I think...

I know...

4,136 → 4,236 → 4,336 → ____

+100 +100 +100

I see ...

I think...

I know...

Questions I have...

____

____

____

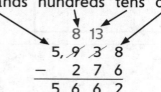

NAME ______________________________ DATE ______________

Lesson 1 Vocabulary Chart

Addition Properties and Subtraction Rules

Use the three-column chart to organize the vocabulary in this lesson. Write the word in Spanish. Then write the correct terms to complete each definition.

English	Spanish	Definition
Commutative Property of Addition		The property that states that the order in which ________ numbers are added ______ _____ change the _____. $12 + 15 = 15 + 12$
Associative Property of Addition		The property that states that the grouping of the addends ______ ____ change the _____. $(4 + 5) + 2 = 4 + (5 + 2)$
Identity Property of Addition		For any number, ______ plus that number is the number. $3 + 0 =$ ___ or $0 + 3 =$ ___
unknown		The amount that has not been ____________.

NAME ______________________________ DATE ______________

Lesson 2 Note Taking

Addition and Subtraction Patterns

Read the question. Write words you need help with and research each word. Use your lesson to write your Cornell notes. Write or draw math examples to explain your thinking. Share your examples with a classmate.

Building on the Essential Question

How can patterns help with addition or subtraction?

Words I need help with:

Notes

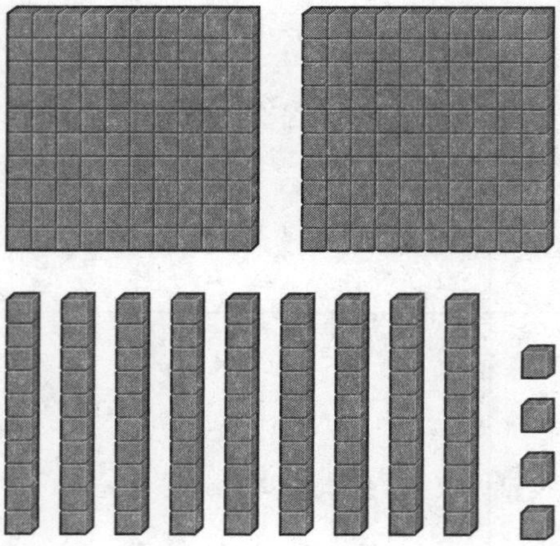

Complete the place value chart for the amount of blocks shown.

hundreds	tens	ones

To find *100 more* than the amount of blocks shown, you use the operation of ____________.

100 more than the amount of blocks shown is ____.

Another 100 more is equal to ____.

To find *10 less* than the amount of blocks shown, you use the operation ____________.

10 less than the amount of blocks shown is ____.

Another 10 less is equal to ____.

My Math Examples:

NAME ________________________________ DATE ______________

Lesson 3 Guided Writing

Add and Subtract Mentally

How do you add and subtract mentally?

Use the exercises below to help you build on answering the Essential Question. Write the correct word or phrase on the lines provided.

1. Rewrite the question in your own words.

____

____

2. What key words do you see in the question?

____

3. If you add ___ to 147, you make 150, which ends in a ______.

4. If you add 3 to one addend, what must you do to the other addend?

____

5. What problem will help you mentally solve 147 + 26? ______________

6. If you subtract ___ from 312, you make 300, which ends in a _________.

7. What problem will help you mentally solve 312 − 241? ______________

8. When you subtract 12 from 312 to help you mentally solve 312 − 241, you must ______ 12 to the difference. 300 − 241 = ___ + 12 = ___

9. Mentally solve 441 − 302. ______

10. How do you add and subtract mentally?

____

____

NAME ______________________________ DATE ______________

Lesson 4 Vocabulary Definition Map

Estimate Sums and Differences

Use the definition map to write a description and list characteristics about the vocabulary word or phrase. Write or draw math examples. Share your examples with a classmate.

My Math Vocabulary:

estimate

Characteristics from Lesson:

5,408 rounded to the nearest ten is ______.

9,214 rounded to the nearest hundred is ______.

3,716 rounded to the nearest thousand is ______.

Description from Glossary:

My Math Examples:

NAME ______________________ DATE ______________

Lesson 5 Concept Web

Add Whole Numbers

Use the concept web to identify parts of finding the sum of whole numbers. Use the terms from the word bank.

Word Bank			
addends	estimate	regroup	sum

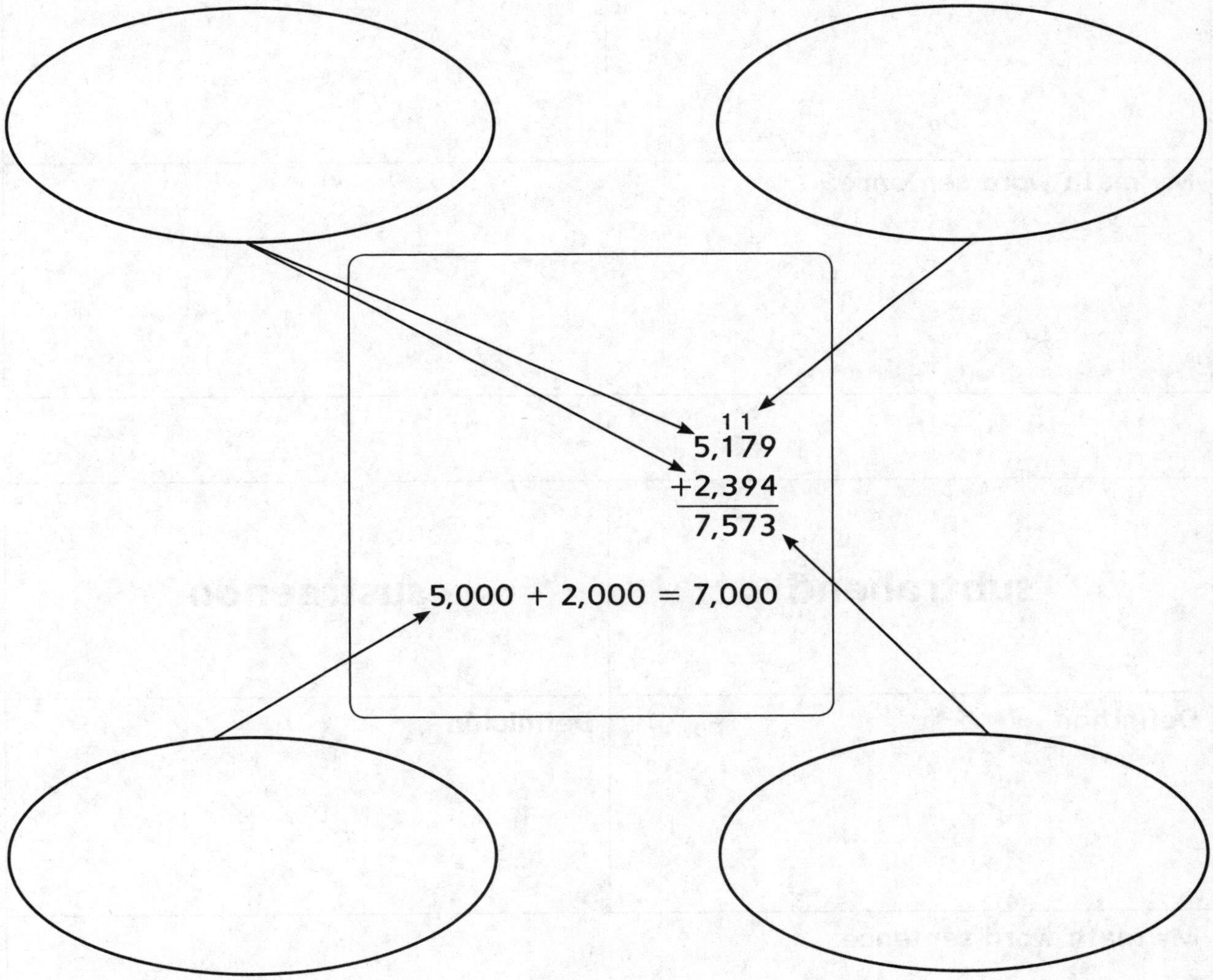

NAME ______________________ DATE ______________

Lesson 6 Vocabulary Cognates

Subtract Whole Numbers

Use the Glossary to define the math word in English and in Spanish in the word boxes. Write a sentence using your math word.

minuend	**minuendo**
Definition	**Definición**
My math word sentence:	

subtrahend	**sustraendo**
Definition	**Definición**
My math word sentence:	

NAME ______________________ DATE ____________

Lesson 7 Note Taking

Subtract Across Zeros

Read the question. Write words you need help with and research each word. Use your lesson to write your Cornell notes. Write or draw math examples to explain your thinking. Share your examples with a classmate.

Building on the Essential Question

How do you subtract across zeros?

Notes:

When you subtract from 2,015, start in the _____ place. Then subtract the _____ place. Then subtract the __________ place, and finally subtract the __________ place.

You can regroup 1 ten as 10 _____.

You can regroup 1 hundred as 10 _____.

You can regroup 1 thousand as 10 __________.

Complete the equation below to find the difference.

```
   □ □
 2 , 0 1 5
−1 , 3 1 4
──────────
 □ , □ □ □
```

Words I need help with:

My Math Examples:

NAME ______________________________ DATE ______________

Lesson 8 Problem-Solving Investigation

STRATEGY: Draw a Diagram

Draw a diagram to help you solve the problems.

1. Twickenham Stadium, in England, can seat **82,000 people.** If there are **49,837 people** seated in the stadium, how many **more people** can be seated in the **stadium?**

Understand	**Solve**
I know: I need to find:	
Plan 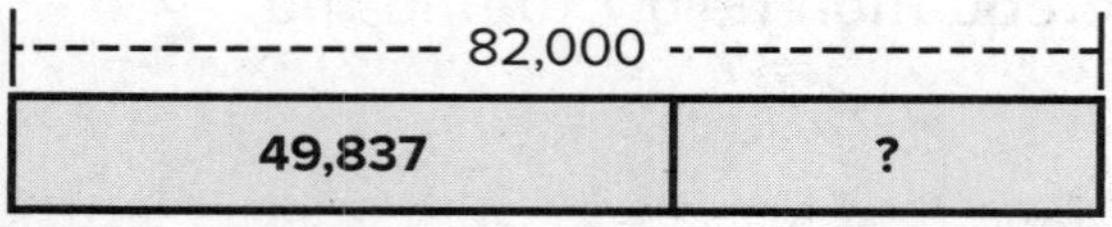	**Check**

2. A bakery uses **ten cups of butter** and ten **eggs** for a recipe. There are **16,280 Calories** in **ten cups of butter**. **Ten eggs** have **1,170 Calories.** **How many more** Calories are there in ten cups of butter **than** in ten eggs?

Understand	**Solve**
I know: I need to find:	
Plan 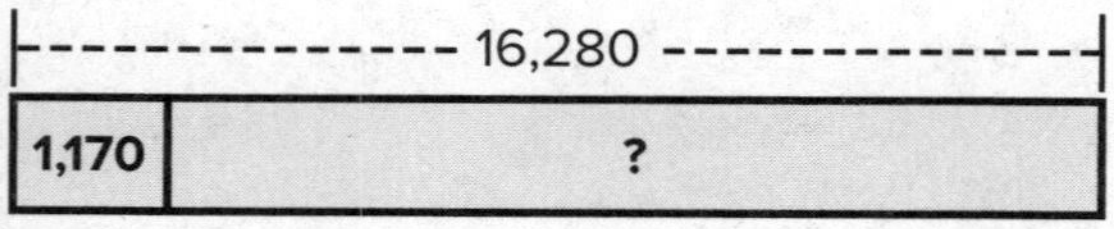	**Check**

NAME ________________________ DATE ____________

Lesson 9 Four-Square Vocabulary

Solve Multi-Step Word Problems

Write the definition for each math word. Write what each word means in your own words. Draw or write examples that show each math word meaning. Then write your own sentences using the words.

Definition	**My Own Words**
equation	
My Examples	**My Sentence**

Definition	**My Own Words**
variable	
My Examples	**My Sentence**

Chapter 3 Understand Multiplication and Division

Inquiry of the Essential Question:

How are multiplication and division related?

Read the Essential Question. Describe your observations (I see...), inferences (I think...), and prior knowledge (I know...) of each math example. Write additional questions you have below. Then share your ideas and questions with a classmate.

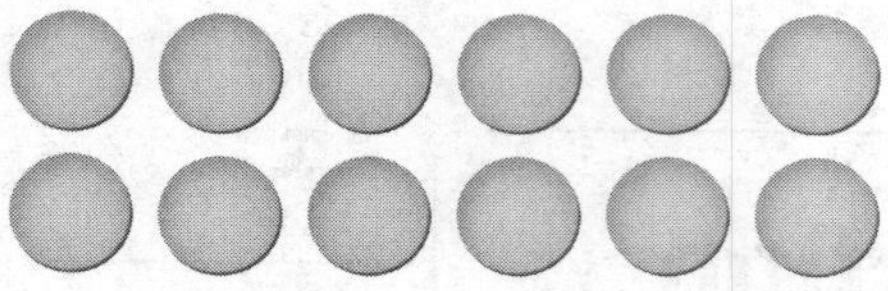

$2 \times 6 = 12$ $\quad 12 \div 2 = 6$

$6 \times 2 = 12$ $\quad 12 \div 6 = 2$

I see ...

I think...

I know...

Words: 5 times more than $4

Diagram:

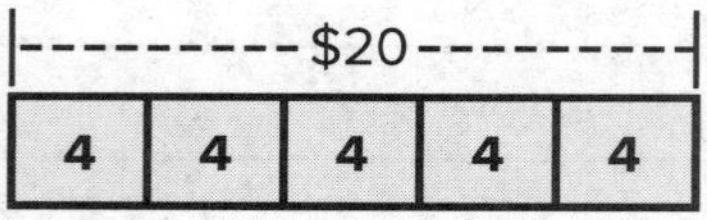

Equation: $5 \times \$4 = \$20$

I see ...

I think...

I know...

$$\begin{array}{r} 30 \\ -\ 6 \\ \hline 24 \end{array} \quad \begin{array}{r} 24 \\ -\ 6 \\ \hline 18 \end{array} \quad \begin{array}{r} 18 \\ -\ 6 \\ \hline 12 \end{array} \quad \begin{array}{r} 12 \\ -\ 6 \\ \hline 6 \end{array} \quad \begin{array}{r} 6 \\ -6 \\ \hline 0 \end{array}$$

I see ...

I think...

I know...

Questions I have...

____

____

____

NAME ______________________________ DATE ______________

Lesson 1 Vocabulary Chart

Relate Multiplication and Division

Use the three-column chart to organize the vocabulary in this lesson. Write the word in Spanish. Then write the correct terms to complete each definition.

English	Spanish	Definition
dividend		A number that is being ________.
divisor		The number by which the ________ is being ________.
factor		A number that ________ a whole number ________. Also a number that is ________ by another number.
product		The answer or result of a ________ problem. It also refers to expressing a number as the ________ of its factors.
quotient		The result of a ________ problem.
fact family		A group of ________ facts using the same ________.

NAME ______________________________ DATE ______________

Lesson 2 Guided Writing

Relate Division and Subtraction

How do you relate division and subtraction?

Use the exercises below to help you build on answering the Essential Question. Write the correct word or phrase on the lines provided.

1. Rewrite the question in your own words.

____

____

2. What key words do you see in the question?

____

3. Repeated addition is ________ the ________ number again and again.

4. The operation of ______________ is the same as repeated ________.

5. Repeated subtraction is ______________ the ______ number again and again.

6. How many times can you subtract 4 from 12 before reaching 0? Use the number line to help you.

____

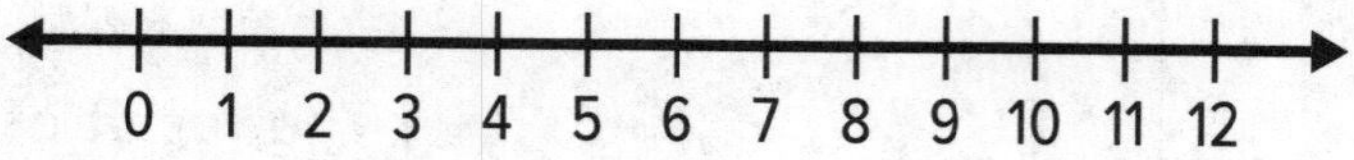

7. What is 12 ÷ 4?

____

8. How do you relate division and subtraction?

____

____

NAME ________________________ DATE ____________

Lesson 3 Concept Web

Multiplication as Comparison

Use the concept web to write examples of ways to indicate 2 × a number.

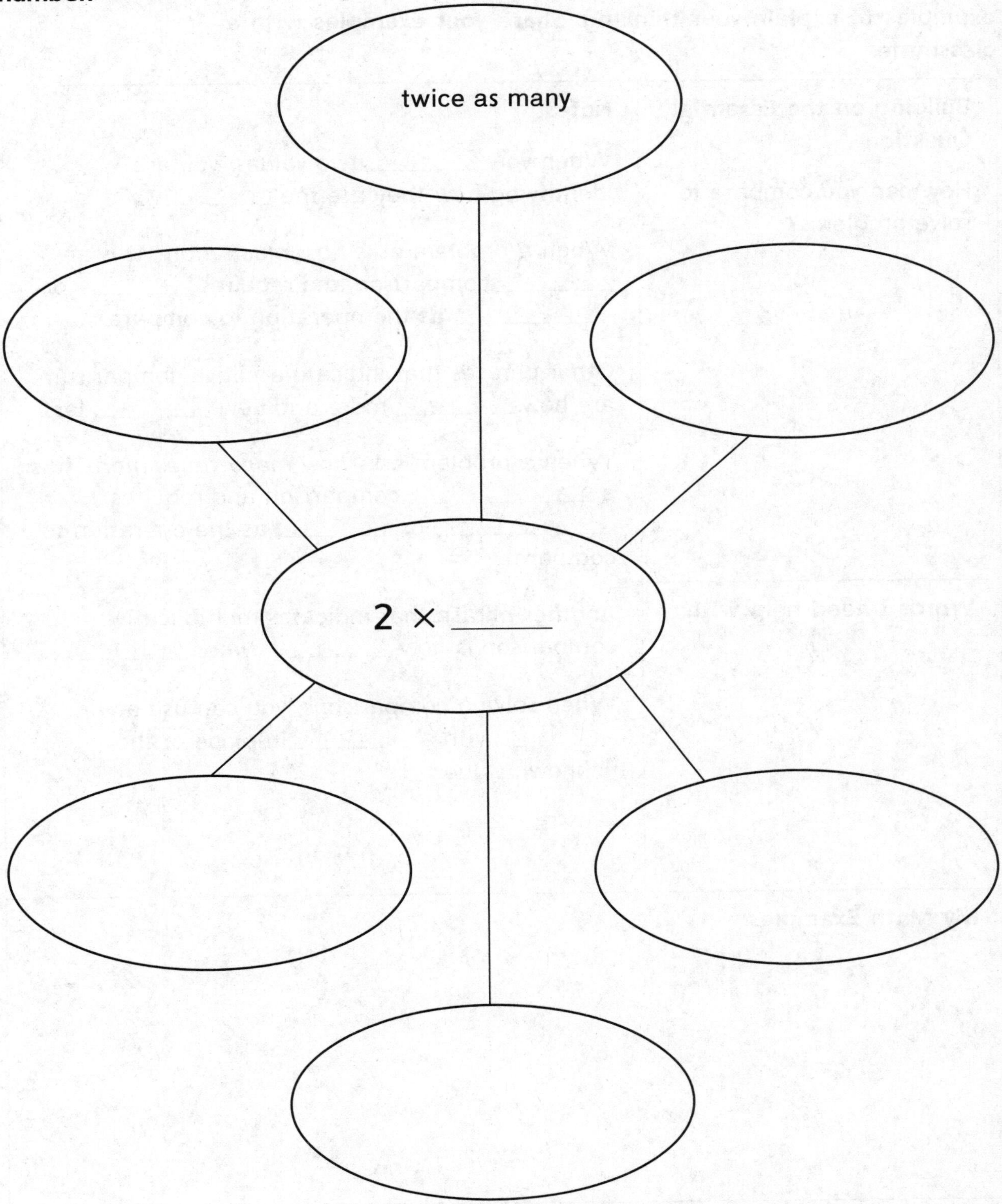

NAME ______________________________ DATE ______________

Lesson 4 Note Taking

Compare to Solve Problems

Read the question. Write words you need help with and research each word. Use your lesson to write your Cornell notes. Write or draw math examples to explain your thinking. Share your examples with a classmate.

Building on the Essential Question

How can you compare to solve problems?

Notes:

When you ________ two values, you are identifying how they are the ________.

When a problem asks *how much more,* it is an ________ comparison and requires ________ or ________ as the operation to compare.

Other phrases that indicate additive comparison are how ________ more and how ________ less.

When a problem asks *how many times more,* it is a ________ comparison and requires ________ or ________ as the operation to compare.

Another phrase that indicates multiplicative comparison is *how* ________ *times* ________.

When solving comparisons, you can use an ________ with a ________ in place of the unknown value.

Words I need help with:

My Math Examples:

NAME ______________________________ DATE ______________

Lesson 5 Vocabulary Cognates

Multiplication Properties and Division Rules

Use the Glossary to define the math word in English and in Spanish in the word boxes. Write a sentence using your math word.

Commutative Property of Multiplication	**propiedad conmutativa de la multiplicación**
Definition	**Definición**
My math word sentence:	

Identity Property of Multiplication	**propiedad de identidad de la multiplicación**
Definition	**Definición**
My math word sentence:	

Zero Property of Multiplication	**propiedad del cero de la multiplicación**
Definition	**Definición**
My math word sentence:	

NAME ______________________________ DATE ______________

Lesson 6 Vocabulary Definition Map

The Associative Property of Multiplication

Use the definition map to write a description and list characteristics about the vocabulary word or phrase. Write or draw math examples. Share your examples with a classmate.

My Math Vocabulary:

Associative Property of Multiplication

Characteristics from Lesson:

The Associative Property of ________ is similar to the Associative Property of Multiplication because they both have __________.

The __________ let you know which two numbers to multiply ____.

This property allows you to determine which ____ factors should be ________ together first to make it easier to find the final ________.

Description from Glossary:

My Math Examples:

NAME ________________________ DATE ____________

Lesson 7 Four-Square Vocabulary

Factors and Multiples

Write the definition for each math word. Write what each word means in your own words. Draw or write examples that show each math word meaning. Then write your own sentences using the words.

Definition	**My Own Words**
My Examples	**My Sentence**

multiple

Definition	**My Own Words**
My Examples	**My Sentence**

decompose

NAME ______________________________ DATE ______________

Lesson 8 Problem-Solving Investigation

STRATEGY: Reasonable Answers

Determine a reasonable answer to each problem.

Pennies Collected	
Child	Number of Pennies
Myron	48
Teresa	52
Veronica	47
Warren	53

1. The **table** shows the number of **pennies collected** by **four children.**
Is it reasonable to say that **Myron** and **Teresa** collected about **100 pennies in** all? Explain.

Understand I know: I need to find:	**Solve**
Plan 48 rounds to... 52 rounds to...	**Check**

2. Jay will make **$240** doing yard work for **6 weeks.**
He (Jay) is **saving** his money to **buy** camping equipment that **costs $400.**
He **has** already saved **$120.**
Is it **reasonable** to say that Jay will **save enough** money **in 6 weeks**? Explain.

Understand I know: I need to find:	**Solve**
Plan  	**Check**

NAME ______________________ DATE ______________

Chapter 4 Multiply with One-Digit Numbers

Inquiry of the Essential Question:

How can I communicate multiplication?

Read the Essential Question. Describe your observations (I see...), inferences (I think...), and prior knowledge (I know...) of each math example. Write additional questions you have below. Then share your ideas and questions with a classmate.

$6 \times 7 = 42$	Basic fact	I see ...
$6 \times 70 = 420$	6×7 tens $= 42$ tens	
$6 \times 700 = 4{,}200$	6×7 hundreds $= 42$ hundreds	I think...
		I know...

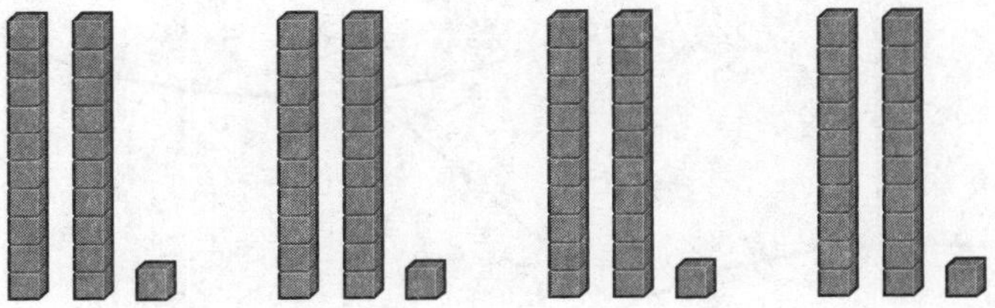

I see ...

I think...

I know...

Number of tens: 8

Number of ones: 4

8 tens + 4 ones = 80 + 4 = 84

$4 \times 21 =$ ______

$$26 \times 7 = (20 \times 7) + (6 \times 7)$$
$$= 140 + 42$$
$$= 182$$

Write 26 as 20 + 6.
Multiply mentally.
Add.

I see ...

I think...

I know...

Questions I have...

____

____

____

NAME ______________________ DATE ______________

Lesson 1 Concept Web

Multiples of 10, 100, and 1,000

Use the concept web to write examples of multiplying using multiples and basic facts.

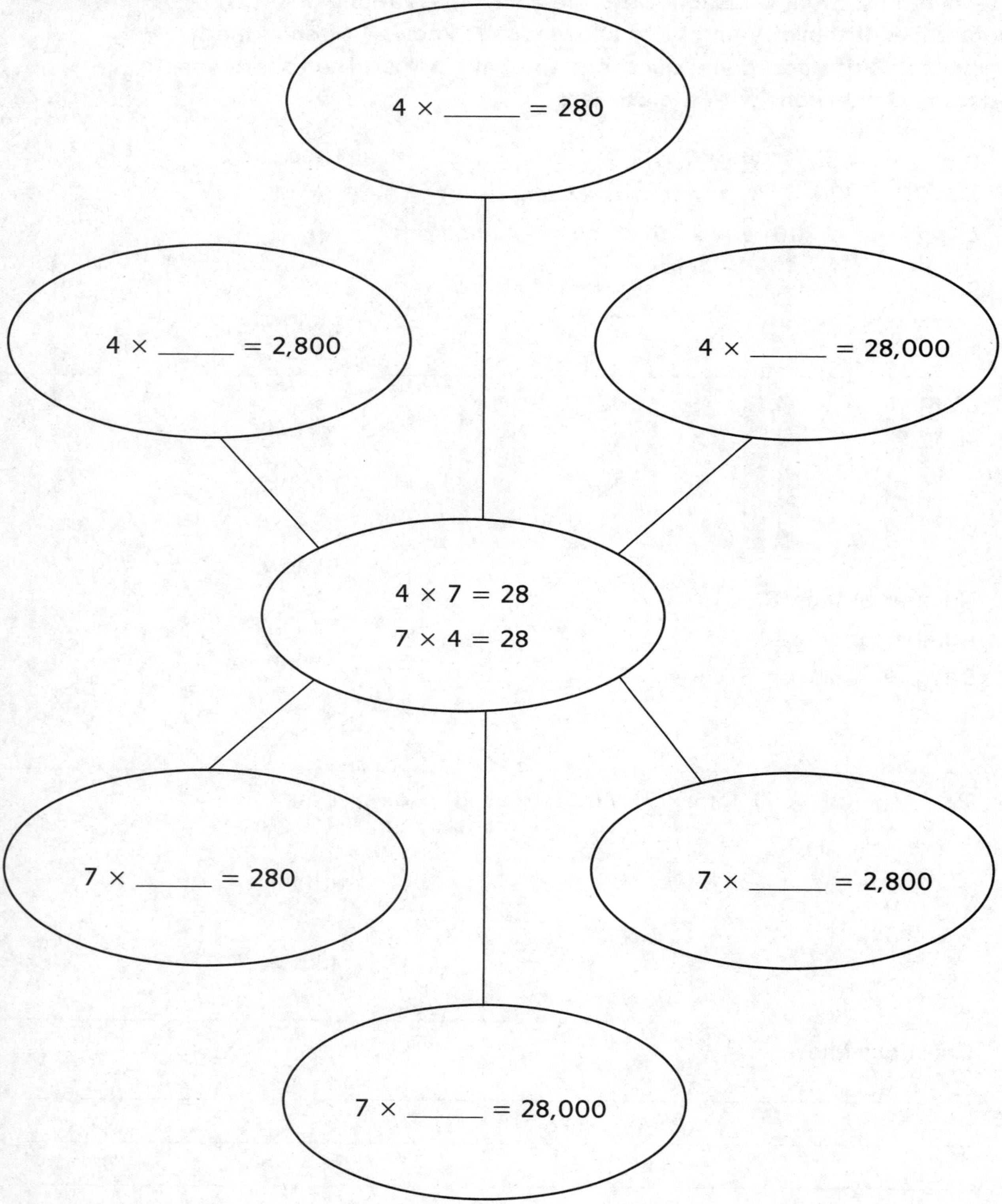

NAME ______________________ DATE ______________

Lesson 2 Multiple Meaning Word

Round to Estimate Products

Complete the four-square chart to review the multiple meaning word or phrase.

Everyday Use	**Math Use in a Sentence**
Math Use	**Example From This Lesson**

round

Complete the sentences below by writing the correct numbers or terms on the lines.

To find the estimate of 917 × 3, round 917 to the greatest place value which is the ____________ place.

Then multiply __________ × 3 to find the estimate __________.

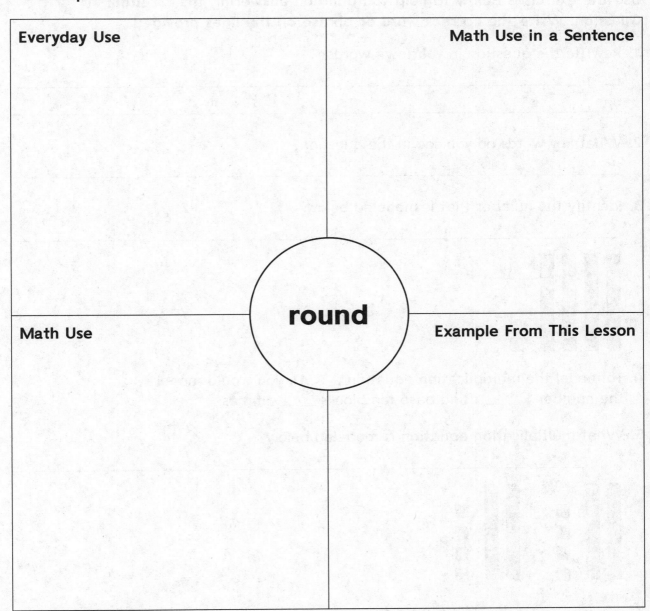

NAME ______________________________ DATE ______________

Lesson 3 Guided Writing

Inquiry/Hands On: Use Place Value to Multiply

How do you use place value to multiply?

Use the exercises below to help you build on answering the Essential Question. Write the correct word or phrase on the lines provided.

1. Rewrite the question in your own words.

____

____

2. What key words do you see in the question?

____

3. Identify the number that is modeled below.

____

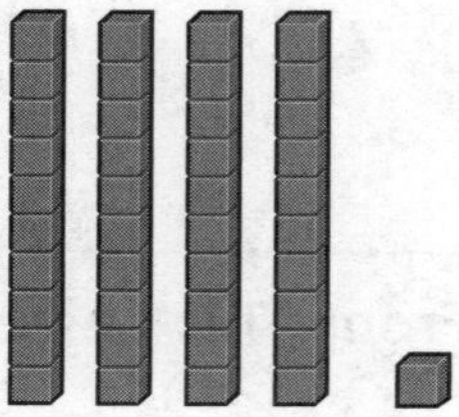

4. To model the multiplication equation 2 × 41, you would model the number ______ using base ten blocks ______ times.

5. What multiplication **equation** is modeled below?

____

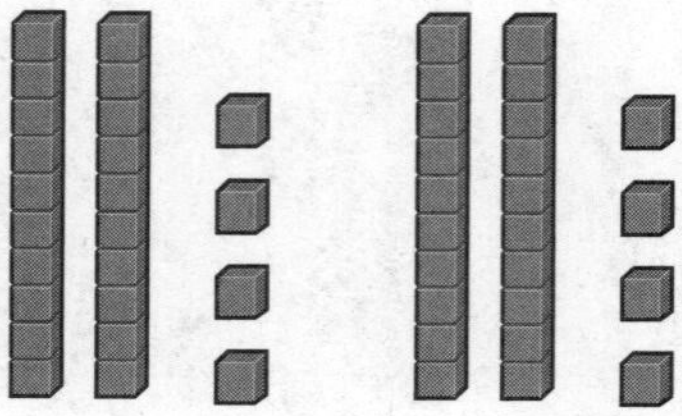

6. Use the modeled multiplication equation above to identify the total number of tens and ones. Record in the place-value chart.

tens	ones

7. How do you use place value to multiply?

____

____

NAME ______________________ DATE ______________

Lesson 4 Four-Square Vocabulary

Inquiry/Hands On: Use Models to Multiply

Write the definition for each math word. Write what each word means in your own words. Draw or write examples that show each math word meaning. Then write your own sentences using the words.

Definition	My Own Words
partial products	
My Examples	My Sentence

Definition	My Own Words
product	
My Examples	My Sentence

NAME ______________________________ DATE ______________

Lesson 5 Note Taking

Multiply by a Two-Digit Number

Read the question. Write words you need help with and research each word. Use your lesson to write your Cornell notes. Write or draw math examples to explain your thinking. Share your examples with a classmate.

Building on the Essential Question	**Notes:**
How do you multiply by a two-digit number?	When you multiply a two-digit number by a one-digit number, first multiply the ________ place of the two-digit number. Then multiply the ________ place of the two-digit number. For the number 34, the tens place value is ____ and the ones place value is ____. For example: 34 $\underline{\times 2}$ First, multiply the ones place. 2 × ___ ones = ___ ones
Words I need help with:	Multiply the tens place. 2 × ___ tens = ___ tens The sum of the partial products is the product. ___ tens and ___ ones is equal to ____. 34 × 2 = ____

My Math Examples:

NAME ______________________ DATE ____________

Lesson 6 Definition Map

Inquiry/Hands On: Model Regrouping

Use the definition map to write a description and list characteristics about the vocabulary word or phrase. Write or draw math examples. Share your examples with a classmate.

My Math Vocabulary:

regroup

Description from Glossary:

Characteristics from Lesson:

You can regroup 10 ______ as 1 ten.

You can regroup 10 ______ as 1 hundred.

You can regroup 10 ____________ as 1 thousand.

My Math Examples:

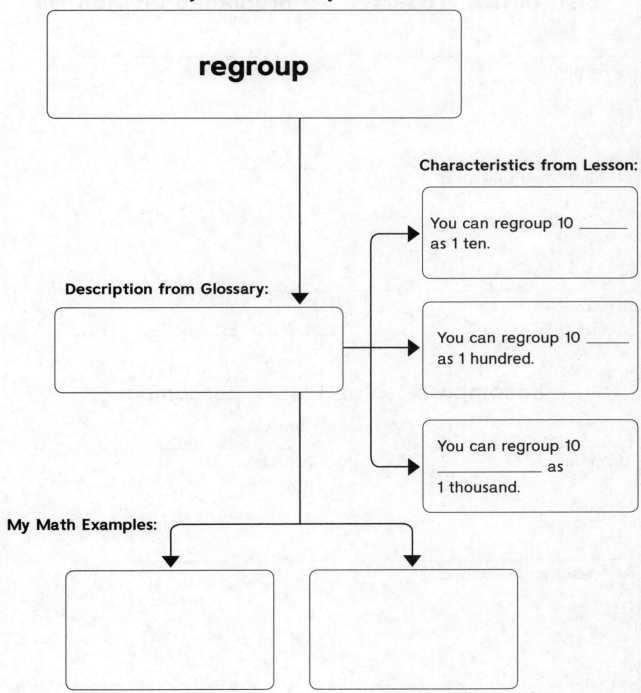

NAME ______________________ DATE ______________

Lesson 7 Vocabulary Cognates

The Distributive Property

Use the Glossary to define the math word in English and in Spanish in the word boxes. Write a sentence using your math word.

Distributive Property	**propiedad distributiva**
Definition	**Definición**
My math word sentence:	

decompose	**descomponer**
Definition	**Definición**
My math word sentence:	

NAME ______________________________ DATE ______________

Lesson 8 Note Taking

Multiply with Regrouping

Read the question. Write words you need help with and research each word. Use your lesson to write your Cornell notes. Write or draw math examples to explain your thinking. Share your examples with a classmate.

Building on the Essential Question

How do you multiply with regrouping?

Words I need help with:

Notes:

When you multiply a two-digit number by a two-digit number, first multiply the ______ place. Then __________ the ones into ________ and ________. Finally multiply the ________ place.

For the number 34, the tens place value is ____ and the ones place value is ____.

For example:

$$\begin{array}{r} 34 \\ \underline{\times 6} \end{array}$$

First, multiply the ones place. 6 × ___ ones = ____ ones.

____ ones are equal to ___ tens and ___ ones.

Next, multiply the tens place. 6 × ___ tens = ____ tens.

Add the tens that were regrouped after multiplying the ones.
____ tens + ___ tens = ____ tens.

The sum of the partial products is the product.
____ tens and ___ ones is equal to _____.

34 × 6 = _____

My Math Examples:

Lesson 9 Concept Web

Multiply by a Multi-Digit Number

Use the concept web to identify each part of the partial product model for 3 × 3,672.

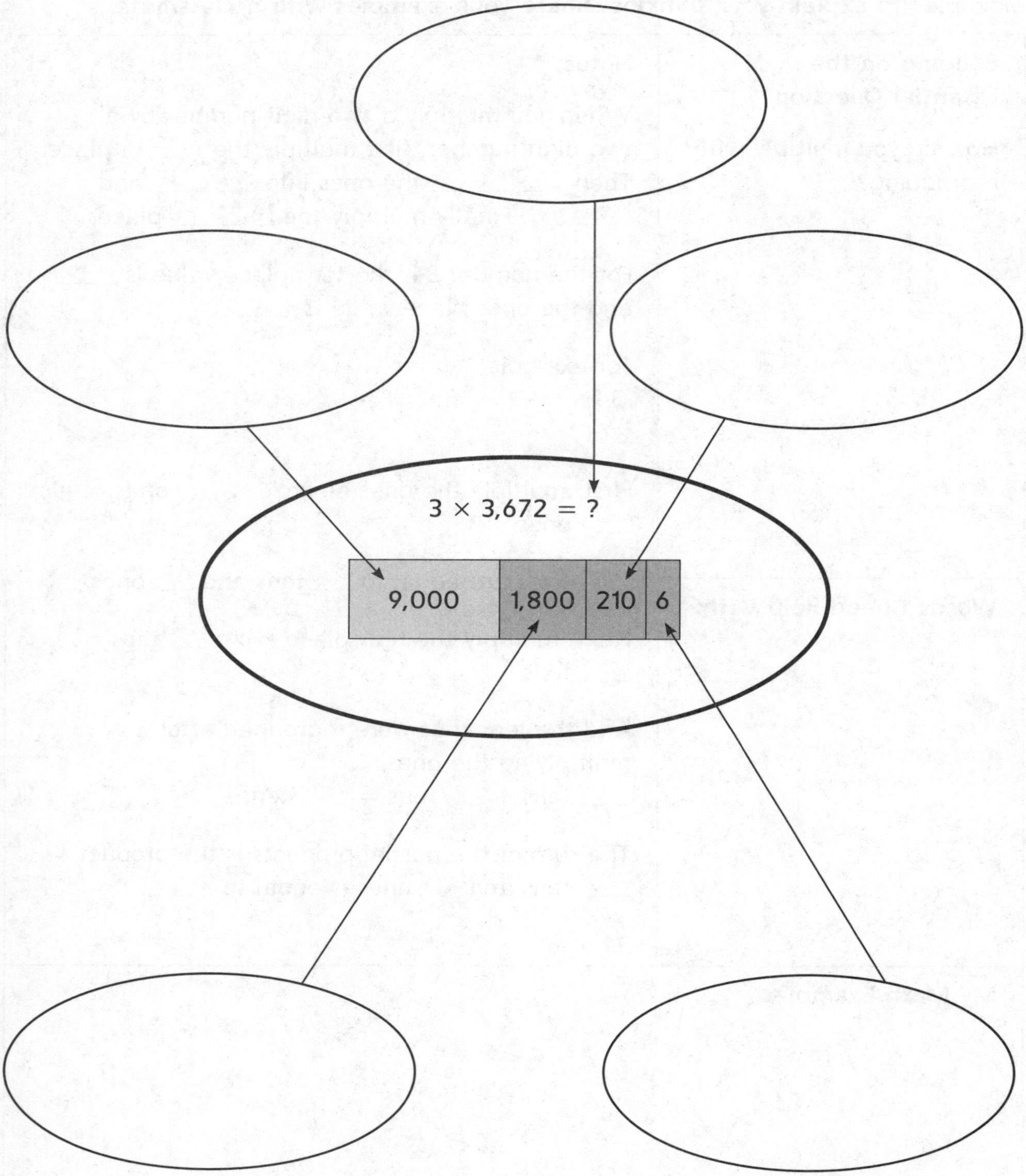

NAME ______________________________ DATE ______________

Lesson 10 Problem-Solving Investigation

STRATEGY: Estimate or Exact Answer

Determine if an estimate or an exact answer is needed to solve each problem.

1. An office needs to buy 6 **computers** and 6 **printers**.
Each **computer** costs $384. Each **printer** costs $88.
About $2,400 will be spent on **computers**.
What is the question?

Understand

Item	Rounded Price (each)	Amount for 6
Computer		
Printer		

Solve
The question is:

Plan

Check

2. **Each** fourth grade **class** reads a total of **495 minutes each week**.
Suppose there are 4 fourth grade **classes**.
How **many minutes** are read **each week?**

Understand

I know:

I need to find:

Solve

Plan
Facts that are important to solve the problem are:

Words that indicate if an exact answer or an estimate is needed are:

Check

NAME ____________________ DATE ____________

Lesson 11 Vocabulary Chart

Multiply Across Zeros

Use the three-column chart to organize the vocabulary in this lesson. Write the word in Spanish. Then write the correct terms to complete each definition.

English	Spanish	Definition
estimate		A number ________ to an exact value. An estimate indicates ________ how much.
multiply		An ____________ on two numbers to find their ____________. It can be thought of as repeated ____________.
partial products		A multiplication method in which the __________ of each place value are found separately, and then ________ together.
Distributive Property		To multiple a ________ by a number, __________ each addend by the number and ______ the products.

NAME ______________________ DATE ______________

Chapter 5 Multiply with Two-Digit Numbers

Inquiry of the Essential Question:

How can I multiply by a two-digit number?

Read the Essential Question. Describe your observations (I see...), inferences (I think...), and prior knowledge (I know...) of each math example. Write additional questions you have below. Then share your ideas and questions with a classmate.

$23 \times 40 = 23 \times (4 \times 10)$	Write 40 as 4×10.	I see ...
$= (23 \times 4) \times 10$	Associative Property of Multiplication	I think...
$= 92 \times 10$	Multiply.	
$= 920$	Use mental math.	I know...

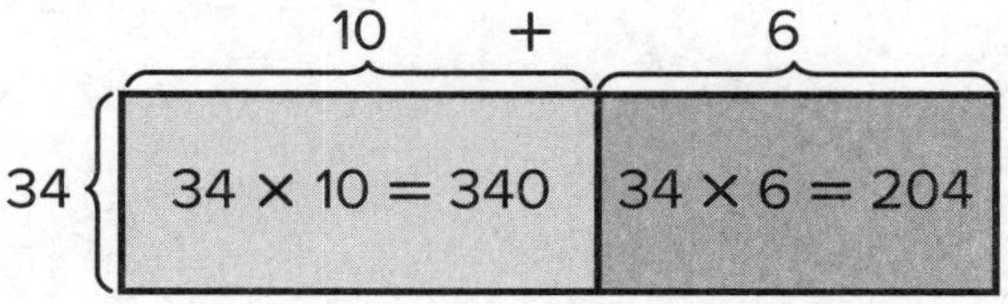

$34 \times 16 = (34 \times 10) + (34 \times 6)$
$= 340 + 204$
$= 544$

I see ...

I think...

I know...

---

56 → rounds to → 60
×37 → rounds to → ×40
2,400

I see ...

I think...

I know...

Questions I have...

____

____

____

NAME ______________________ DATE ______________

Lesson 1 Concept Web

Multiply by Tens

Use the concept web to write examples of multiples of 10.

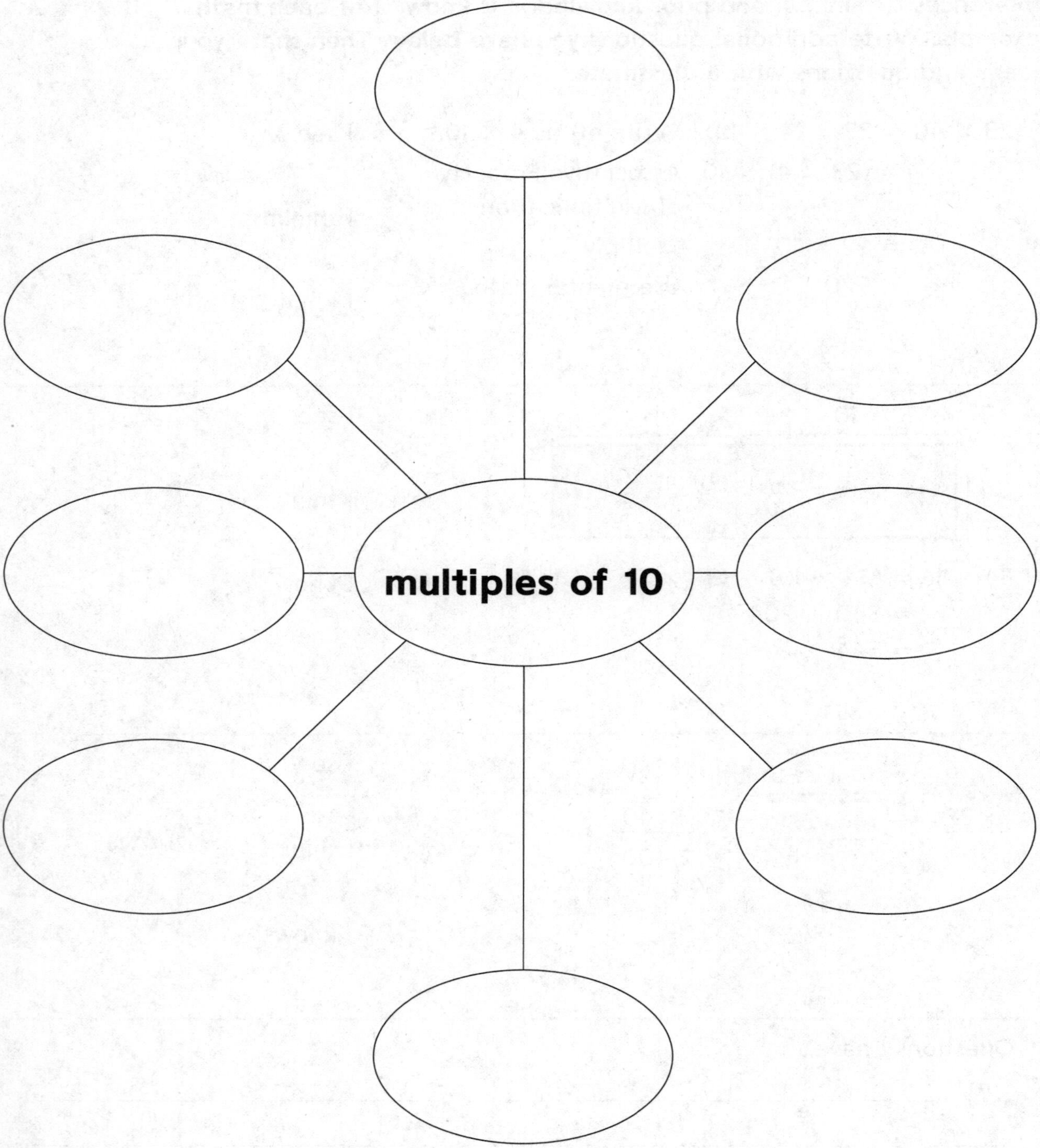

NAME ______________________ DATE ______________

Lesson 2 Guided Writing

Estimate Products

How do you estimate products?

Use the exercises below to help you build on answering the Essential Question. Write the correct word or phrase on the lines provided.

1. Rewrite the question in your own words.

   ____

   ____

2. What key words do you see in the question?

   ____

3. When you ________ a number, you make it easier to work with by changing the ________ of the number.

4. To ________ a number to the nearest ten, look at the ________ place.

5. If the value in the ________ place is greater than 5, round ______ to the nearest ten.

6. If the value in the ones place is less than 5, round ________ to the __________ ten.

7. An __________ is a number close to an exact value. An estimate indicates ________ how much.

8. How do you estimate products?

   ____

   ____

NAME ______________________ DATE ______________

Lesson 3 Vocabulary Definition Map

Inquiry/Hands On: Use the Distributive Property to Multiply

Use the definition map to write a description and list characteristics about the vocabulary word or phrase. Write or draw math examples. Share your examples with a classmate.

My Math Vocabulary:

Distributive Property of Multiplication

Description from Glossary:

Characteristics from Lesson:

A two-digit number can be decomposed into the sum of the values of the ______ place and the ______ place.

The ______________ let you know that you should do that operation first.

$4 \times 12 = 4 \times (\_\_ + \_\_)$

$= (4 \times \_\_) + (4 \times \_\_)$

$= \_\_ + \_\_$

$= \_\_$

My Math Examples:

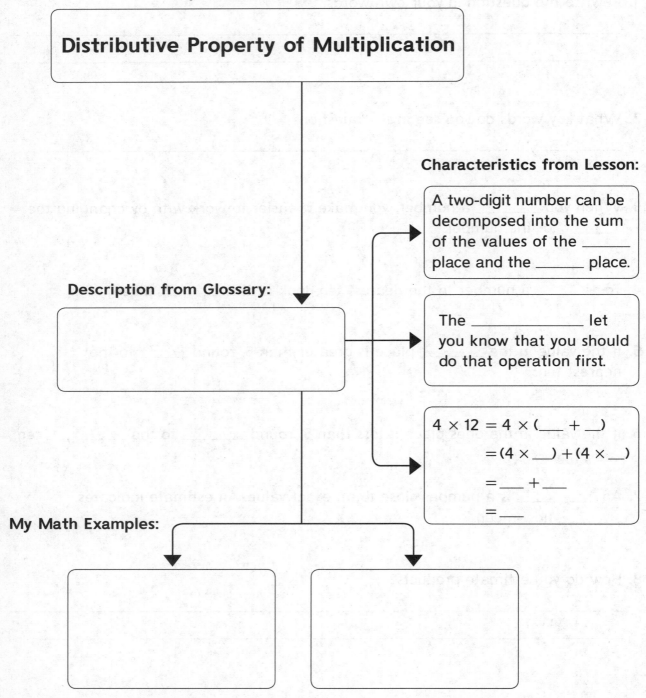

NAME ______________________ DATE ______________

Lesson 4 Vocabulary Cognates

Multiply by a Two-Digit Number

Use the Glossary to define the math word in English and in Spanish in the word boxes. Write a sentence using your math word.

product	**producto**
Definition	**Definición**
My math word sentence:	

partial products	**productos parciales**
Definition	**Definición**
My math word sentence:	

NAME ______________________________ DATE ______________

Lesson 5 Note Taking

Solve Multi-Step Word Problems

Read the question. Write words you need help with and research each word. Use your lesson to write your Cornell notes. Write or draw math examples to explain your thinking. Share your examples with a classmate.

Building on the Essential Question How do you solve multi-step word problems?	**Notes:** An operation is a mathematical process such as ____________ (+), ____________ (−), ____________ (×), or ____________ (÷). A ____________ is a letter or symbol used to represent an ____________ quantity. ____________ ( ) are the enclosing symbols which indicate that the terms within are a unit. When you see the words: *less than, fewer, remains*, or *difference* in a word problem, the operation to use is most likely ____________. When you see the words: *total, altogether, both,* or *sum* in a word problem, the operation to use is most likely ____________. When you see the words: *times, each, every day, at this rate,* or *product* in a word problem, the operation to use is most likely ____________.
Words I need help with:	

My Math Examples:

NAME ________________________________ DATE ______________

Lesson 6 Problem-Solving Investigation

STRATEGY: Make a Table

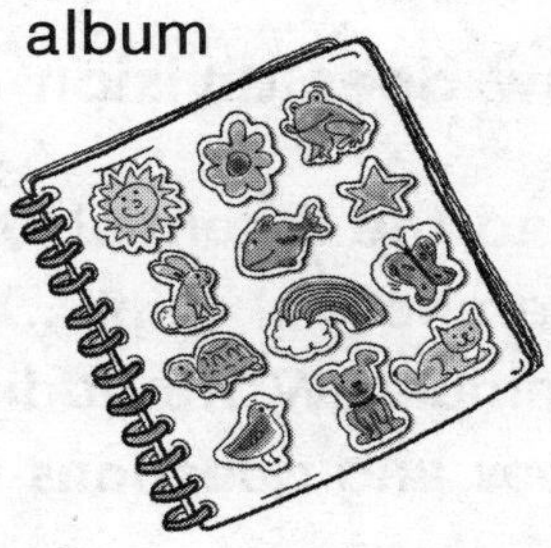

Make a table to solve each problem.

1. **A page** from **Dana's** album is shown.
Dana puts the **same** number of stickers on **each** page.
She (Dana) has **30 pages** of stickers.
How many stickers does she have **in all**?

Understand

I know:

I need to find:

Solve

Pages	1	10	20	30
Stickers	12			

Dana has ______ stickers.

Plan

I will make a __________ to solve.

Check

2. West Glenn School has **23 students** in **each** class.
There are 6 fourth grade **classes**.
About **how many** fourth grade **students** are there **in all**?

Understand

I know:

I need to find:

Solve

Classes	1	2	3	4	5	6
Students	20					

There are **about** ______ students.

Plan

I will make a __________ to solve.

Check

NAME ______________________________ DATE ______________

Chapter 6 Divide by a One-Digit Number

Inquiry of the Essential Question:

How does division affect numbers?

Read the Essential Question. Describe your observations (I see...), inferences (I think...), and prior knowledge (I know...) of each math example. Write additional questions you have below. Then share your ideas and questions with a classmate.

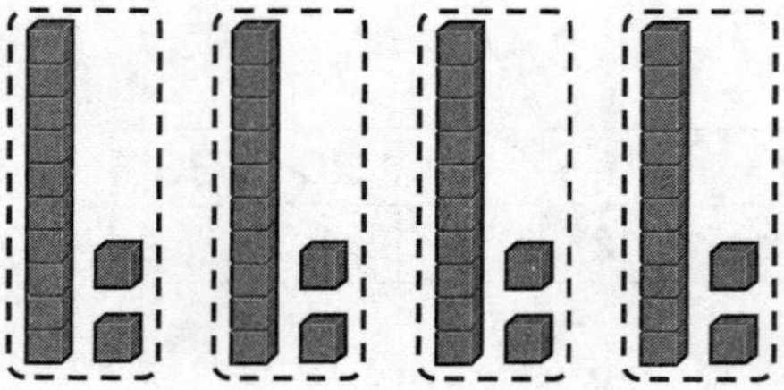

There is 1 ten and 2 ones, or 12 in each group, so $48 \div 4 = 12$.

I see ...

I think...

I know...

$$\begin{array}{r} 5\text{ R1} \\ 5\overline{)26} \\ -25 \\ \hline 1 \end{array}$$

Divide
Multiply
Subtract

I see ...

I think...

I know...

$36 \div 6 = 6$ | $6 \times 6 = 36$
$360 \div 6 = 60$ | $6 \times 60 = 360$
$3{,}600 \div 6 = ?$ | $6 \times 600 = 3{,}600$

So, $3{,}600 \div 6 = 600$

I see ...

I think...

I know...

Questions I have...

____

____

____

NAME ______________________ DATE ______________

Lesson 1 Vocabulary Cognates

Divide Multiples of 10, 100, and 1,000

Use the Glossary to define the math word in English and in Spanish in the word boxes. Write a sentence using your math word.

dividend	**dividendo**
Definition	**Definición**
My math word sentence:	

multiple	**múltiplo**
Definition	**Definición**
My math word sentence:	

NAME ______________________________ DATE ______________

Lesson 2 Vocabulary Definition Map

Estimate Quotients

Use the definition map to write a description and list characteristics about the vocabulary word or phrase. Write or draw math examples. Share your examples with a classmate.

My Math Vocabulary:

compatible numbers

Characteristics from Lesson:

Compatible numbers can be used to ________ quotients.

Sometimes you round down to find the compatible number. The compatible numbers for estimating 453 ÷ 7 are ____ ÷ 7 = 60.

Sometimes you round up to find the compatible number. The compatible numbers for estimating 453 ÷ 8 are ____ ÷ 8 = 60.

Description from Glossary:

My Math Examples:

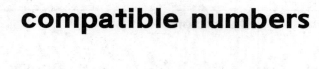

NAME ______________________ DATE ______________

Lesson 3 Concept Web

Inquiry/Hands On: Use Place Value to Divide

Use the concept web to write each part of a division equation.

Word Bank

dividend divisor remainder quotient

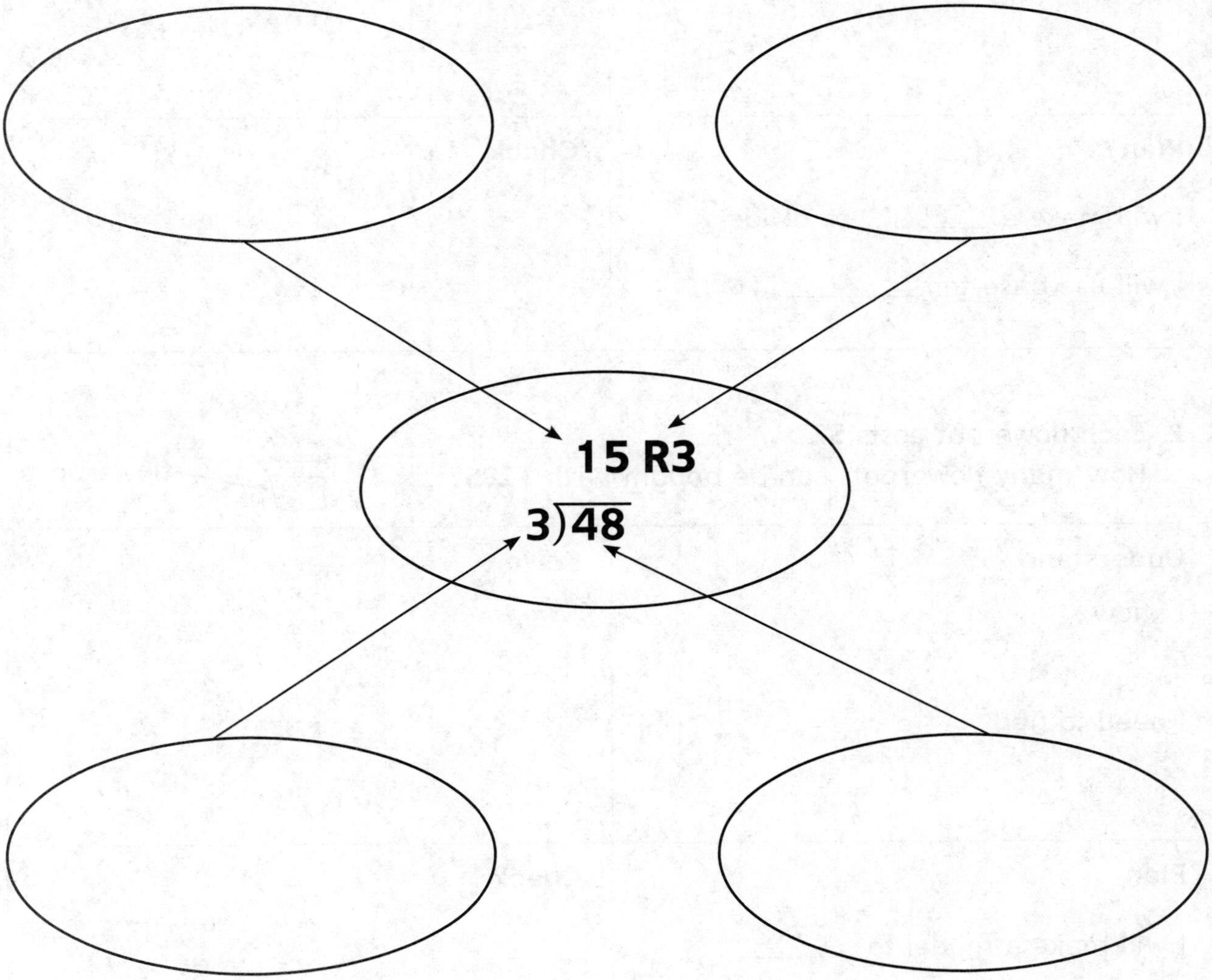

NAME ______________________________ DATE ______________

Lesson 4 Problem-Solving Investigation

STRATEGY: Make a Model

Make a model to help you solve each problem.

1. **Casey's <u>mom</u>** is the baseball **<u>coach</u>** for his team. She (mom/coach) **spent $150** on baseballs. **<u>Each</u>** baseball **cost $5.** **How many** baseballs did **she buy**?

baseball

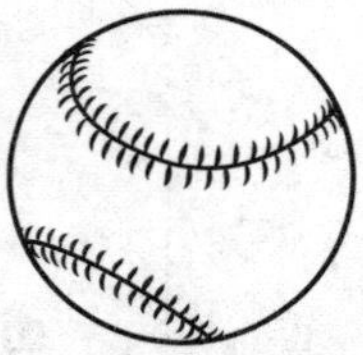

Understand I know: I need to find:	**Solve**
Plan I will make a ________ to divide. I will use base-ten ________.	**Check**

2. **Each** flowerpot costs **$7.** **How many** flowerpots can be bought **with $285**?

flowerpot

Understand I know: I need to find:	**Solve**
Plan I will make a model to ________. I will use ________ blocks.	**Check**

NAME ______________________________ DATE ______________

Lesson 5 Note Taking

Divide with Remainders

Read the question. Write words you need help with and research each word. Use your lesson to write your Cornell notes. Write or draw math examples to explain your thinking. Share your examples with a classmate.

Building on the Essential Question

How can you divide with remainders?

Notes:

Steps to ________ a two-digit number by a one-digit number.

1. Divide the ______

- multiply
- __________
- compare
- bring ______ the ones

2. Divide the ______

- __________
- subtract
- __________
- bring down the ______

Words I need help with:

If there are no __________ to bring down, the number left over is called the __________.

```
   19 R1
2)39
 -2↓
  19
 -18
   1 ← number left over
```

My Math Examples:

NAME ________________________ DATE ____________

Lesson 6 Multiple Meaning Word

Interpret Remainders

Complete the four-square chart to review the multiple meaning word.

Everyday Use	**Math Use in a Sentence**
remainder	
Math Use	**Example From This Lesson**

Write the correct terms on the lines to complete the sentences.

When you divide 153 by 10, you will have a quotient of ___ with a remainder of ___.

153 people were riding vans to a park. Each van holds 10 people each. How many vans would be needed to move all 153 people? ___

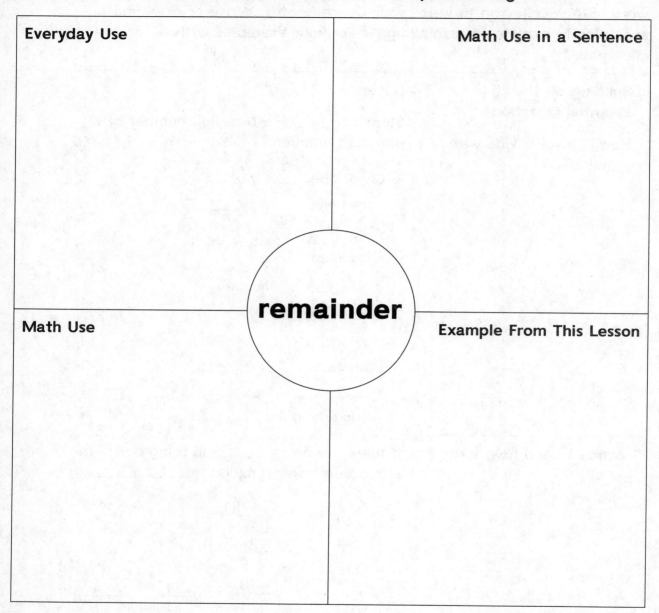

NAME ______________________________ DATE ______________

Lesson 7 Note Taking

Place the First Digit

Read the question. Write words you need help with and research each word. Use your lesson to write your Cornell notes. Write or draw math examples to explain your thinking. Share your examples with a classmate.

Building on the Essential Question

How can you place the first digit?

Notes:

Sometimes the first digit of the ________ is less than the divisor. You may not be able to _____ the first digit of the quotient over the first digit of the dividend.

$3\overline{)28}$

Since 3 is greater than 2, you cannot divide the tens.

Begin division with the _____.
Divide the ones place. ___ ones ÷ 3 = __ ones

Place the 9 over the __ in the ones place. → 9 R1

$$\begin{array}{r} 9\text{ R}1 \\ 3\overline{)28} \\ -27 \\ \hline 1 \end{array}$$

Next, multiply: 3 × __ ones = ___ ones
Subtract the 27 from 28.
___ ones − ___ ones = __ one.
Compare to the divisor.
Since 1 is _____ than 3, 1 is the remainder.

28 ÷ 3 = __ R __

Words I need help with:

My Math Examples:

NAME ______________________________ DATE ______________

Lesson 8 Guided Writing

Inquiry/Hands On: Distributive Property and Partial Quotients

How do you divide using the distributive property and partial quotients?

Use the exercises below to help you build on answering the Essential Question. Write the correct word or phrase on the lines provided.

1. Rewrite the question in your own words.

____

____

2. What key words do you see in the question?

____

3. When you ____________ a number, you break the number into different parts.

4. Decompose 648 into different parts according to place value.

____

5. *Partial quotients* is a ________ method in which the ________ is separated into sections that are easy to divide.

6. Complete the equations that will give the partial quotients for the division of 693 ÷ 3.

____ ÷ 3 = ____
___ ÷ 3 = ___
__ ÷ 3 = __

7. Find the sum of the partial quotients and solve for 693 ÷ 3.

____ + ___ + __ = ____
693 ÷ 3 = ____

8. How do you use the distributive property and partial quotients to divide?

____

____

NAME ______________________ DATE ____________

Lesson 9 Concept Web

Divide Greater Numbers

Use the concept web to identify the place value of each digit.

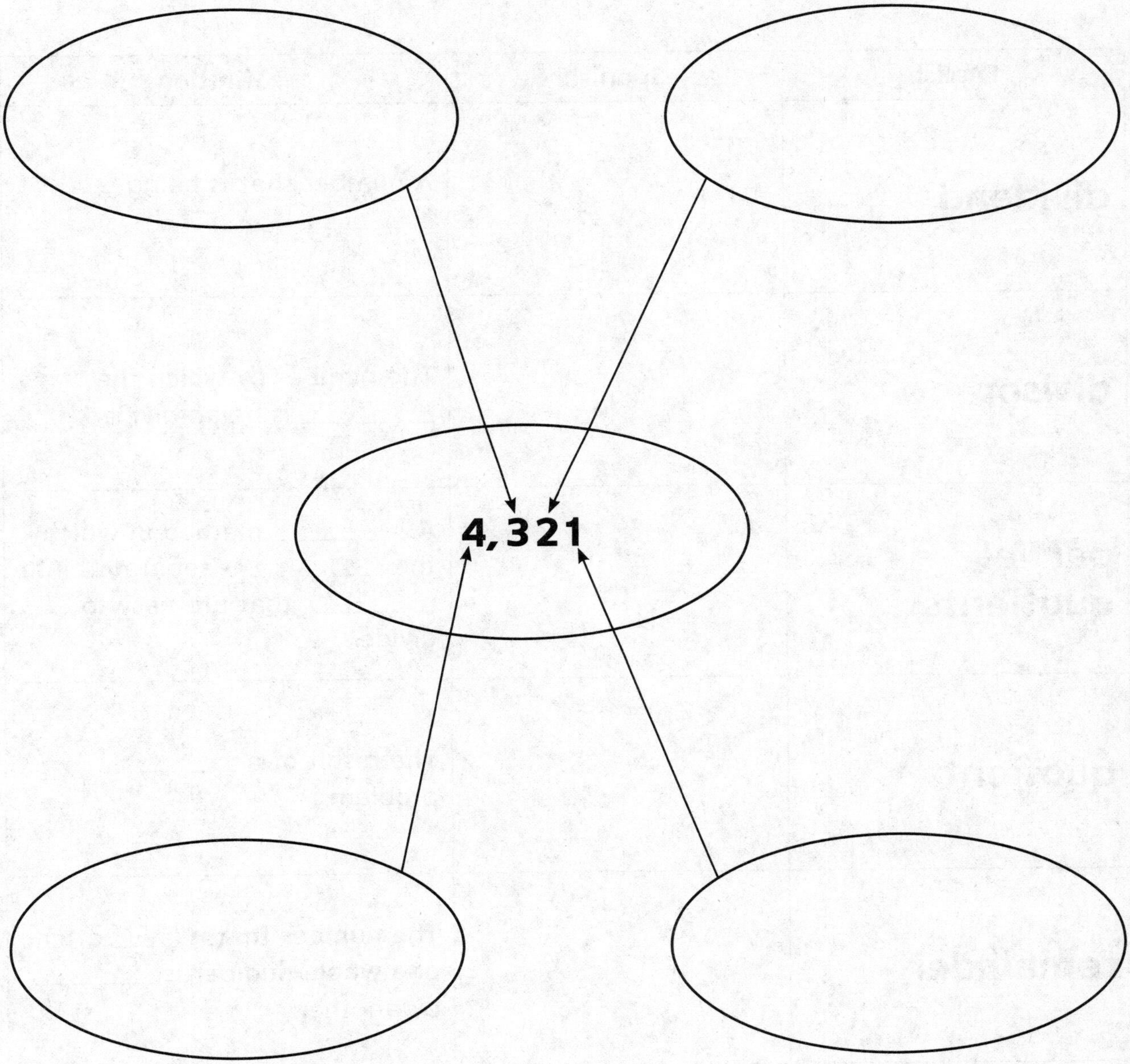

NAME ______________________ DATE ______________

Lesson 10 Vocabulary Chart

Quotients with Zeros

Use the three-column chart to organize the review vocabulary in this lesson. Write the word in Spanish. Then write the correct terms to complete each definition.

English	Spanish	Definition
dividend		A number that is being ________.
divisor		The number by which the ________ is being divided.
partial quotients		A ________ method in which the ________ is separated into ________ that are easy to divide.
quotient		The result of a ________ problem.
remainder		The number that is _____ after one whole number is ________ by another.

NAME ______________________ DATE ____________

Lesson 11 Four-Square Vocabulary

Solve Multi-Step Word Problems

Write the definition for each math word. Write what each word means in your own words. Draw or write examples that show each math word meaning. Then write your own sentences using the words.

Definition	**My Own Words**
equation	
My Examples	**My Sentence**

Definition	**My Own Words**
parentheses	
My Examples	**My Sentence**

NAME ______________________________ DATE ______________

Chapter 7 Patterns and Sequences

Inquiry of the Essential Question:

How are patterns used in mathematics?

Read the Essential Question. Describe your observations (I see...), inferences (I think...), and prior knowledge (I know...) of each math example. Write additional questions you have below. Then share your ideas and questions with a classmate.

55, 56, 52, 53, 49, 50, ___?___

+1 −4 +1 −4 +1

The pattern is add 1, then subtract 4.

I see ...

I think...

I know...

Input (a)	Output (b)
4	12
7	21
10	30
13	?

Rule: Multiply by 3.
Equation: $a \times 3 = b$

I see ...

I think...

I know...

If $x = 6$, what is the value of y in $2 \times (9 + x) = y$?

$2 \times (9 + x) = y$
$2 \times (9 + 6) = y$
$2 \times 15 = y$
$30 = y$

I see ...

I think...

I know...

Questions I have...

____

____

____

NAME ______________________________ DATE ______________

Lesson 1 Vocabulary Definition Map

Nonnumeric Patterns

Use the definition map to write a description and list characteristics about the vocabulary word or phrase. Write or draw math examples. Share your examples with a classmate.

My Math Vocabulary:

nonnumeric pattern

Description from Glossary:

Characteristics from Lesson:

A pattern is a sequence of numbers, figures, or symbols that follows a ______ or design.

"Non" means "not", so nonnumeric means ______ ________.

The nonnumeric pattern shown below is ______, ______, ______.

My Math Examples:

NAME ______________________ DATE ______________

Lesson 2 Concept Web

Numeric Patterns

Use the given rule to write the next number in each numeric pattern shown in the concept web.

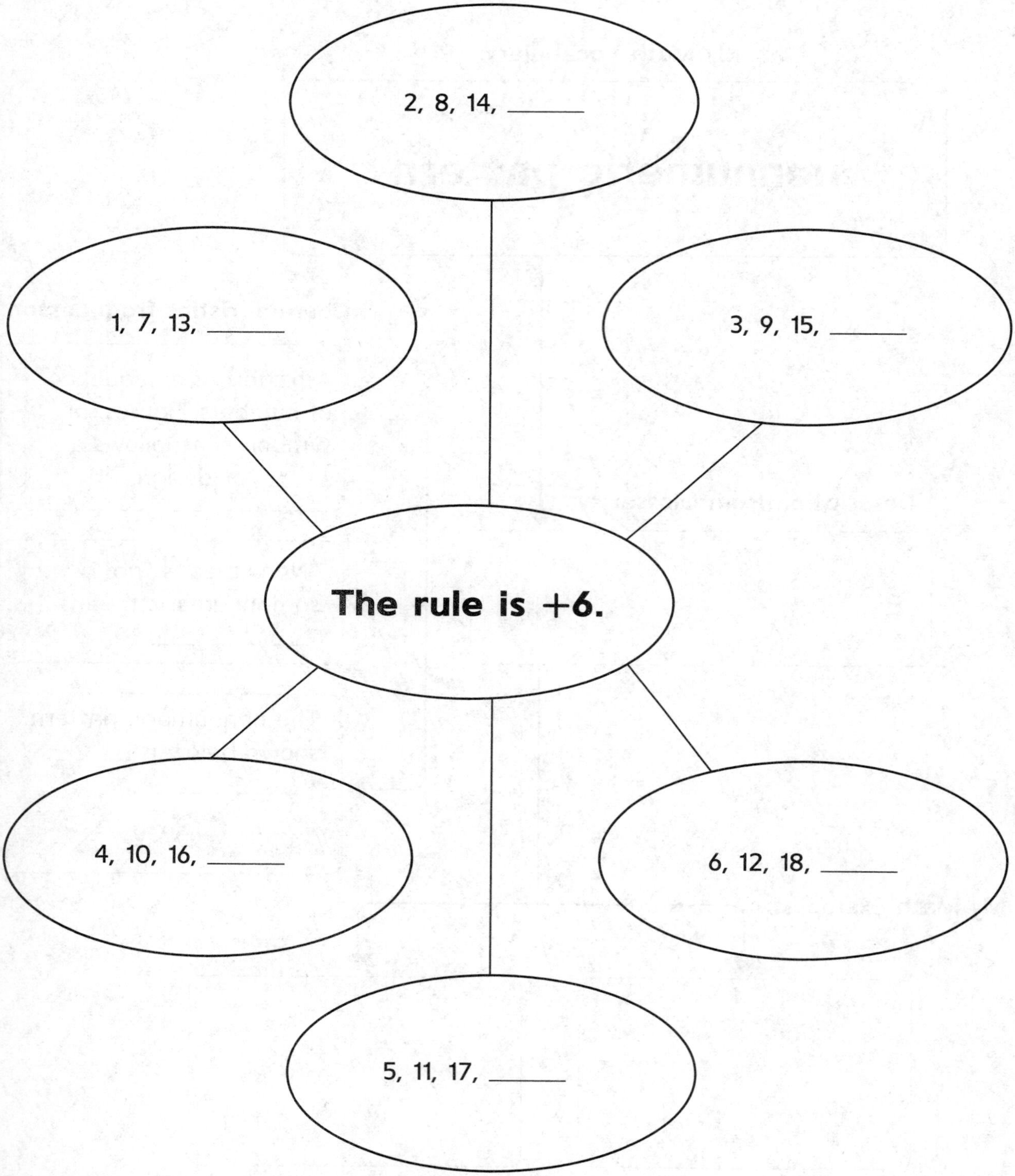

NAME ______________________ DATE ______________

Lesson 3 Note Taking

Sequences

Read the question. Write words you need help with and research each word. Use your lesson to write your Cornell notes. Write or draw math examples to explain your thinking. Share your examples with a classmate.

<table>
<tr>
<td>

Building on the Essential Question

How are sequences used in mathematics?

</td>
<td rowspan="2">

Notes:

A term is each number in a ________ ________.

A sequence is an ordered arrangement of ______ that make up a ________.

A rule is a statement that describes a __________ between numbers or objects.

The number 8 is the first _____ in the sequence 8, 15, 22, 29, 36.

I can observe the _________ to find the rule.

The rule is add ____.

I know to use the rule on the last term of the sequence to ______ the pattern. 36 + 7 = ____

I can extend the pattern to find the next three terms: 8, 15, 22, 29, 36, ____, ____, ____.

</td>
</tr>
<tr>
<td>

Words I need help with:

</td>
</tr>
<tr>
<td colspan="2">

My Math Examples:

</td>
</tr>
</table>

NAME ______________________________ DATE ______________

Lesson 4 Problem-Solving Investigation

STRATEGY: Look for a Pattern

Look for a pattern to solve each problem.

airplane

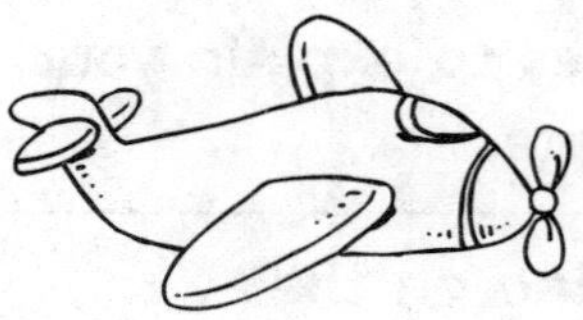

1. A store sold **48** model airplanes in **August**, **58** model airplanes in **September**, and **68** model airplanes in **October**. Suppose this **pattern continues**. **How many** model airplanes will be sold in **December**?

Understand I know: I need to find:	**Solve** Sequence: Rule: Next term in pattern:
Plan I will look for a __________ to solve the problem.	**Check**

2. The **table** shows how many tickets were **sold** for the school play **each day**. Based on the **pattern**, how many tickets will be **sold** on **Friday**?

Day	Number of Tickets
Monday	312
Tuesday	316
Wednesday	320
Thursday	324

Understand I know: I need to find:	**Solve** Sequence: Rule: Next term in pattern:
Plan I will look for a __________ to solve the problem.	**Check**

NAME ______________________ DATE ______________

Lesson 5 Note Taking

Addition and Subtraction Rules

Read the question. Write words you need help with and research each word. Use your lesson to write your Cornell notes. Write or draw math examples to explain your thinking. Share your examples with a classmate.

Building on the Essential Question

How are addition and subtraction rules like patterns?

Notes:

The rule in the pattern below is ____________.
150, 142, 134, 126, 118

To find the next term, ____________ from the last term. 118 − ____ = ____

An *input* is a quantity that is changed to produce an ________.

An *output* is the result of an input quantity being ________.

Input *(x)*	Output *(y)*
118	110
110	

Words I need help with:

The first input *(x)* is ____.
The first output *(y)* is ____.
I know I must subtract 8 from 118 to get ____.

The second input *(x)* is ____.
If I subtract 8 from the input *(x)* 110, I get an output of ____.

The numbers change in the same way each time.
The rule is $x - \_\_\_\_ = y$.

My Math Examples:

NAME ______________________ DATE ____________

Lesson 6 Four-Square Vocabulary

Multiplication and Division Rules

Write the definition for each review math word. Write what each word means in your own words. Draw or write examples that show each math word meaning. Then write your own sentences using the words.

Definition	**My Own Words**
division	
My Examples	**My Sentence**

Definition	**My Own Words**
multiplication	
My Examples	**My Sentence**

NAME ______________________________ DATE ______________

Lesson 7 Concept Web

Order of Operations

Use the concept web to identify the order to perform the operations in the equation.

Word Bank			
first	second	third	fourth

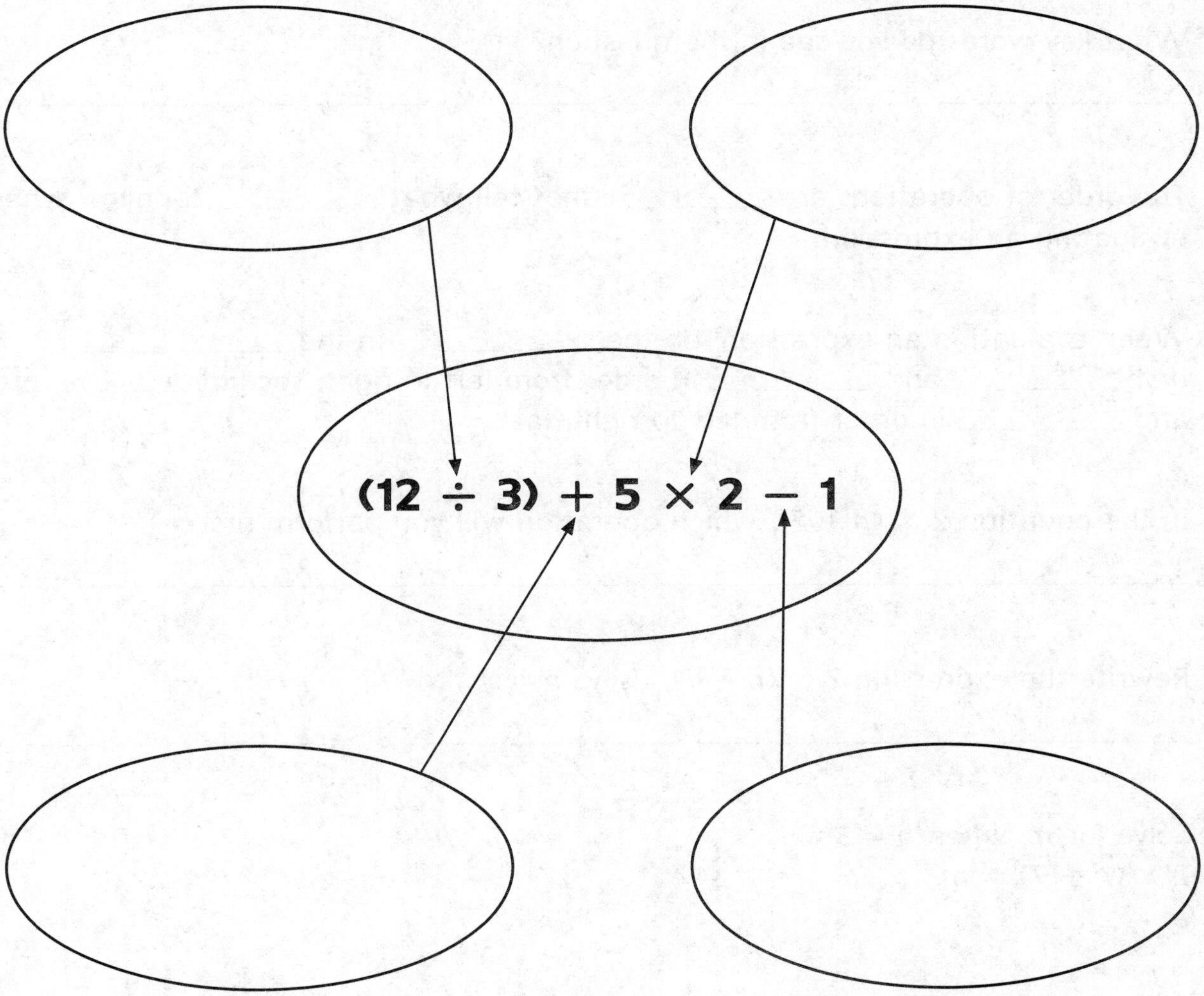

NAME ______________________________ DATE ______________

Lesson 8 Guided Writing

Inquiry/Hands On: Equations with Two Operations

How do you solve equations with two operations?

Use the exercises below to help you build on answering the Essential Question. Write the correct word or phrase on the lines provided.

1. Rewrite the question in your own words.

____

____

2. What key words do you see in the question?

____

3. The order of operations are __________ that tell what __________ to follow when evaluating an expression.

4. When evaluating an expression, do the __________ in the __________ first. __________ and __________ in order from left to right, second. __________ and __________ in order from left to right, last.

5. In the equation $2 \times (n + 7)$, which operation will you perform first?

____

6. Rewrite the expression $2 \times (n + 7)$, using $n = 3$.

____

7. Solve for m, when $n = 3$.
$2 \times (n + 7) = m$.
$m =$ ______

8. How do you solve equations with two operations?

____

____

NAME ________________________________ DATE ______________

Lesson 9 Multiple Meaning Word

Equations with Multiple Operations

Complete the four-square chart to review the multiple meaning word or phrase.

Everyday Use	Math Use in a Sentence
operation	
Math Use	Example From This Lesson

Write the correct symbol, or term on each line to complete the sentence.

A family bought three pizzas that cost $11 each. They had a coupon for $5 off any order. The equation that gives the total cost of the order uses two operations. The equation is $11 ____ 3 ____ $5 = $28. The two operations are ________________ and ________________.

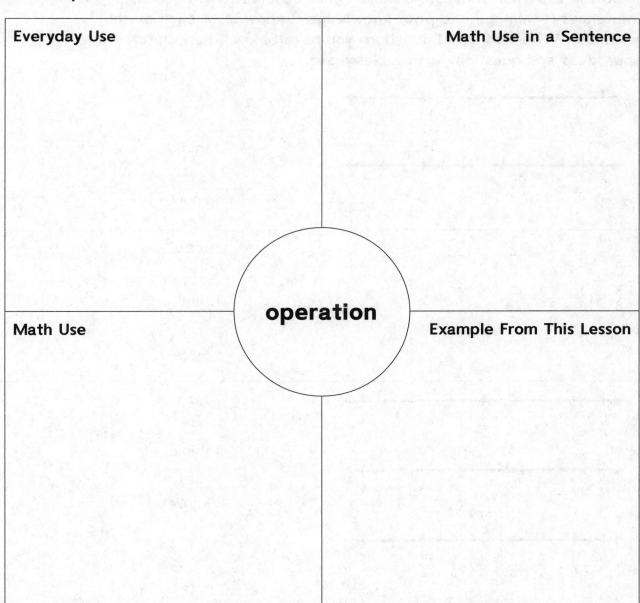

NAME ______________________ DATE ____________

Chapter 8 Fractions

Inquiry of the Essential Question:

How can different fractions name the same amount?

Read the Essential Question. Describe your observations (I see...), inferences (I think...), and prior knowledge (I know...) of each math example. Write additional questions you have below. Then share your ideas and questions with a classmate.

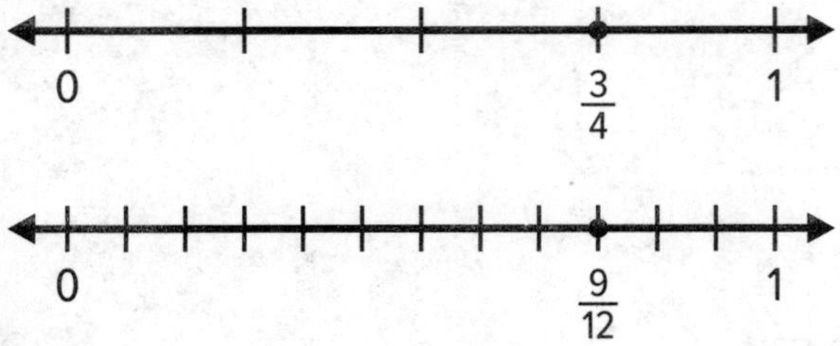

I see ...

I think...

I know...

$\frac{3}{8} = \frac{3 \times 2}{8 \times 2} = \frac{6}{16}$

$\frac{3}{8} = \frac{3 \times 3}{8 \times 3} = \frac{9}{24}$

I see ...

I think...

I know...

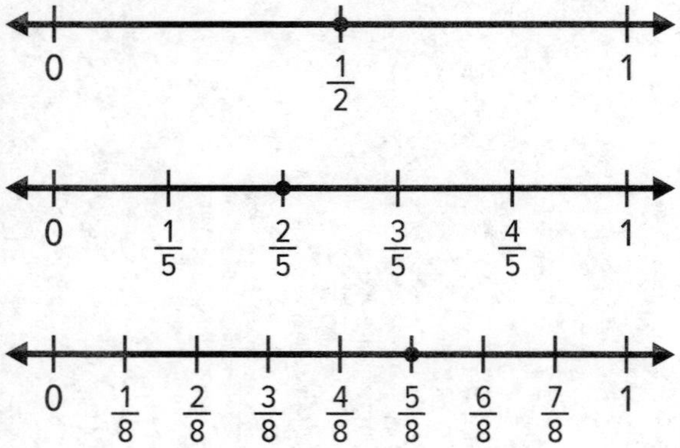

$\frac{2}{5} < \frac{1}{2}$ and $\frac{5}{8} > \frac{1}{2}$. So, $\frac{2}{5} < \frac{5}{8}$.

I see ...

I think...

I know...

Questions I have...

____

____

____

NAME ______________________ DATE ______________

Lesson 1 Vocabulary Definition Map

Factors and Multiples

Use the definition map to write a description and list characteristics about the vocabulary word or phrase. Write or draw math examples. Share your examples with a classmate.

My Math Vocabulary:

factor pairs

Characteristics from Lesson:

A factor is a number that divides a whole number __________.

A factor is a number that is __________ by another number.

A number, like 12, can have multiple factor pairs. For example:

$1 \times 12 = 12$
$2 \times \_\_\_\_ = 12$
$3 \times \_\_\_\_ = 12$

Description from Glossary:

My Math Examples:

NAME ______________________ DATE ______________

Lesson 2 Vocabulary Cognates

Prime and Composite Numbers

Use the Glossary to define the math word in English and in Spanish in the word boxes. Write a sentence using your math word.

prime number	**número primo**
Definition	**Definición**
My math word sentence:	

composite number	**número compuesto**
Definition	**Definición**
My math word sentence:	

NAME ____ DATE ____________________

Lesson 3 Vocabulary Chart

Inquiry/Hands On: Model Equivalent Fractions

Use the three-column chart to organize the vocabulary in this lesson. Write the word in Spanish. Then write the correct terms to complete each definition.

English	Spanish	Definition
denominator		The ________ number in a fraction.
equivalent fractions		Fractions that represent the __________ number.
numerator		The number __________ the bar in a fraction. The part of the fraction that tells _____ ______ of the equal parts are being ________.
fraction		A number that represents part of a __________ or part of a ______.

NAME ____________________ DATE ____________

Lesson 4 Concept Web

Equivalent Fractions

Use the concept web to identify equivalent fractions. Write "true" if the fractions are equivalent and "false" if they are not.

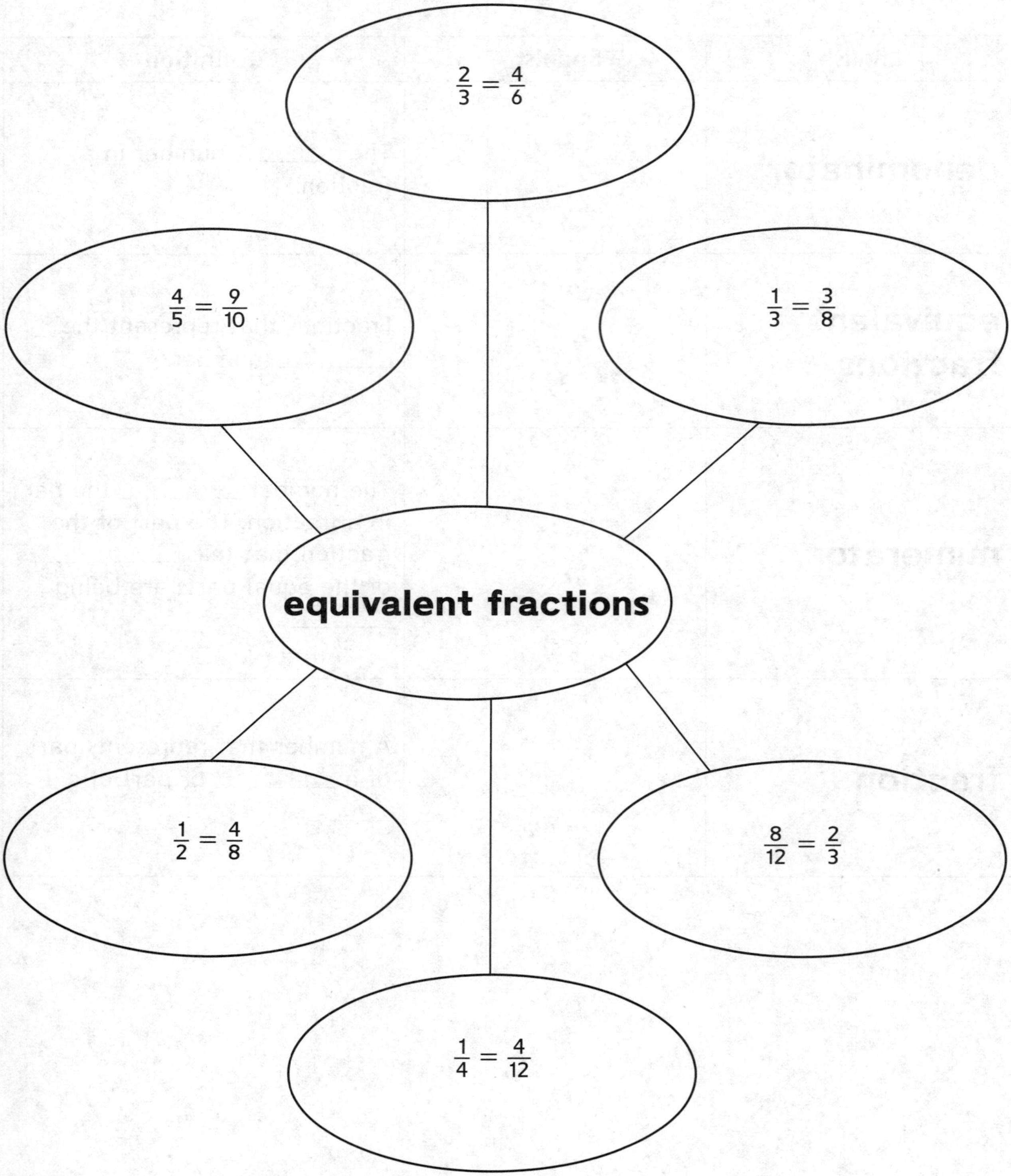

NAME ______________________________ DATE ______________

Lesson 5 Four-Square Vocabulary

Simplest Form

Write the definition for each math word. Write what each word means in your own words. Draw or write examples that show each math word meaning. Then write your own sentences using the words.

Definition	**My Own Words**
My Examples	**My Sentence**

greatest common factor

Definition	**My Own Words**
My Examples	**My Sentence**

simplest form

NAME ______________________________ DATE ______________

Lesson 6 Note Taking

Compare and Order Fractions

Read the question. Write words you need help with and research each word. Use your lesson to write your Cornell notes. Write or draw math examples to explain your thinking. Share your examples with a classmate.

Building on the Essential Question How do I compare and order fractions?	**Notes:** In the fraction $\frac{5}{8}$, the numerator is ___ and the denominator is ___. To compare the fractions $\frac{2}{3}$ and $\frac{5}{8}$, you must first find equivalent fractions, so that the ______________ are the same. Find the least common multiple before finding equivalent fractions. The multiples of 3 are: 3, 6, 9, 12, ___, ___, ___, ◯. The multiples of 8 are: 8, 16, (24), 32, ___, ___, ___, ___. The least common multiple of 3 and 8 is ___.
Words I need help with:	Next, find equivalent fractions with a denominator of 24. $\frac{2}{3} = \frac{\quad}{24}$ and $\frac{5}{8} = \frac{\quad}{24}$ Compare the fractions. $\frac{\quad}{24} > \frac{\quad}{24}$, so $\frac{2}{3}$ ◯ $\frac{5}{8}$.

My Math Examples:

NAME ______________________________ DATE ______________

Lesson 7 Guided Writing

Use Benchmark Fractions to Compare and Order

How do you use benchmark fractions to compare and order?

Use the exercises below to help you build on answering the Essential Question. Write the correct word or phrase on the lines provided.

1. Rewrite the question in your own words.

____

____

2. What key words do you see in the question?

____

3. A ________ is a number that represents part of a whole or part of a set.

4. Benchmark fractions are ________ fractions that are used for ________.

5. Compare. $\frac{1}{2} \bigcirc \frac{3}{8}$

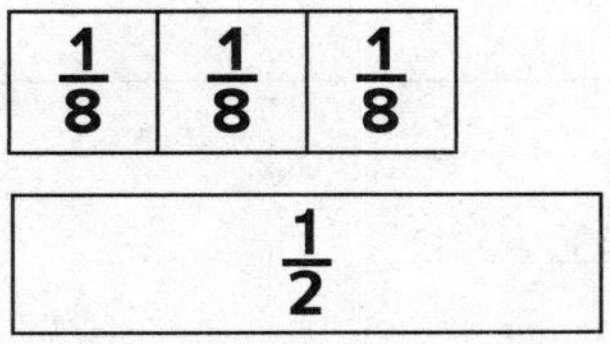

6. Compare. $\frac{1}{2} \bigcirc \frac{3}{5}$

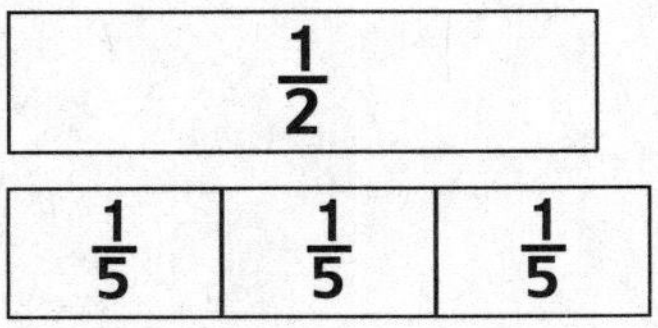

7. Compare. $\frac{3}{8} \bigcirc \frac{3}{5}$

8. How do you use benchmark fractions to compare and order?

____

____

NAME ______________________________ DATE ______________

Lesson 8 Problem-Solving Investigation

STRATEGY: Use Logical Reasoning

Use logical reasoning to solve each problem.

1. Sophia is making a salad with **tomatoes**, **cucumbers**, and **mozzarella cheese**.
Use the clues to find the **amount** of **each ingredient**.
The amounts are $\frac{3}{6}$ cup, $\frac{2}{5}$ cup, and $\frac{3}{4}$ cup.
There is **less** amount of tomatoes **than** cucumbers.
There is **less** amount of cheese **than** tomatoes.

Understand I know: I need to find:	Solve
Plan	**Check**

2. **Mason** walked on **Monday**, **Wednesday**, and **Friday**.
Use the clues to find **how far** he (Mason) walked **each day**.
The distances were $\frac{6}{8}$ mile, $\frac{1}{4}$ mile, and $\frac{1}{6}$ mile.
He did **not** walk the **farthest** on Monday.
He walked **less** on Friday **than** Monday.

Understand I know: I need to find:	Solve
Plan	**Check**

NAME ______________________________ DATE ______________

Lesson 9 Vocabulary Definition Map

Mixed Numbers

Use the definition map to write a description and list characteristics about the vocabulary word or phrase. Write or draw math examples. Share your examples with a classmate.

My Math Vocabulary:

mixed number

Characteristics from Lesson:

________ __________ are the numbers 0, 1, 2, 3, 4, ...

A __________ is a number that represents part of a whole or part of a set.

A mixed number represents an amount ________ than 1.

Description from Glossary:

My Math Examples:

NAME ____ DATE ____________________

Lesson 10 Note Taking

Mixed Numbers and Improper Fractions

Read the question. Write words you need help with and research each word. Use your lesson to write your Cornell notes. Write or draw math examples to explain your thinking. Share your examples with a classmate.

Building on the Essential Question

How do I convert between mixed numbers and improper fractions?

Notes:

A fraction with a numerator that is greater than or equal to the denominator is an __________ __________.

A number that has a whole number part and a fraction part is a __________ __________.

An improper fraction can be written as a __________ number.

A mixed number can be written as an __________ fraction.

These are equivalent fractions for one whole.

$$1 = \frac{1}{1} = \frac{2}{2} = \frac{\ }{3} = \frac{\ }{4} = \frac{\ }{5} = \frac{\ }{6}$$

I can use the equivalent fractions of one whole to write a mixed number as an improper fraction.

$$1\frac{1}{3} = 1 + \frac{1}{3} = \frac{\ }{3} + \frac{1}{3} = \frac{\ }{3}$$

I can use the equivalent fractions of one whole to write an improper fraction as a mixed number.

$$\frac{8}{5} = \frac{5}{5} + \frac{\ }{5} = 1 + \frac{\ }{5} = 1\frac{\ }{5}$$

Words I need help with:

My Math Examples:

NAME ______________________________ DATE ______________

Chapter 9 Operations with Fractions

Inquiry of the Essential Question:

How can I use operations to model real-world fractions?

Read the Essential Question. Describe your observations (I see...), inferences (I think...), and prior knowledge (I know...) of each math example. Write additional questions you have below. Then share your ideas and questions with a classmate.

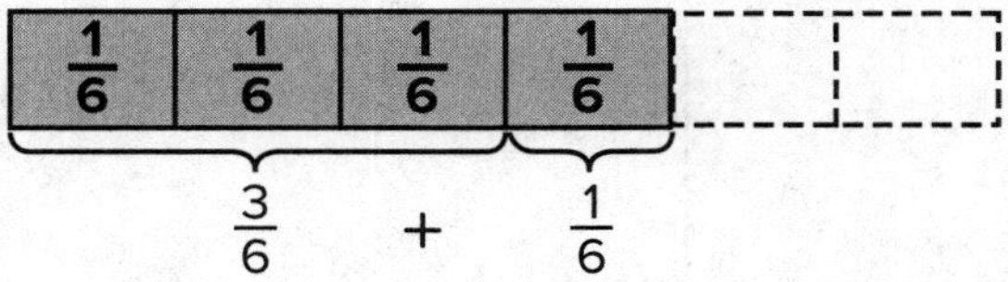

$\frac{3}{6} + \frac{1}{6} = \frac{4}{6}$

I see ...

I think...

I know...

$3\frac{2}{5} = \frac{5}{5} + \frac{5}{5} + \frac{5}{5} + \frac{1}{5} + \frac{1}{5}$

$= \frac{5 + 5 + 5 + 1 + 1}{5} = \frac{17}{5}$

$1\frac{1}{5} = \frac{5}{5} + \frac{1}{5} = \frac{5 + 1}{5} = \frac{6}{5}$

$\frac{17}{5} - \frac{6}{5} = \frac{11}{5}$ or $2\frac{1}{5}$

I see ...

I think...

I know...

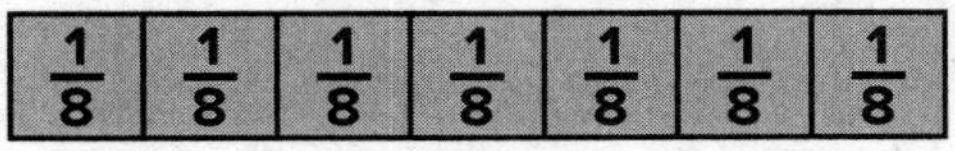

$\frac{7}{8} = 7 \times \frac{1}{8}$

I see ...

I think...

I know...

Questions I have...

____

____

____

NAME ______________________________ DATE ______________

Lesson 1 Vocabulary Definition Map

Inquiry/Hands On: Use Models to Add Like Fractions

Use the definition map to write a description and list characteristics about the vocabulary word or phrase. Write or draw math examples. Share your examples with a classmate.

My Math Vocabulary:

like fractions

Characteristics from Lesson:

A denominator is the _______ number in a fraction.

The word "like" indicates that the fractions are of the _____ form. The form is the denominator.

When adding like fractions, the _____ has the same denominator as the addends.

Description from Glossary:

My Math Examples:

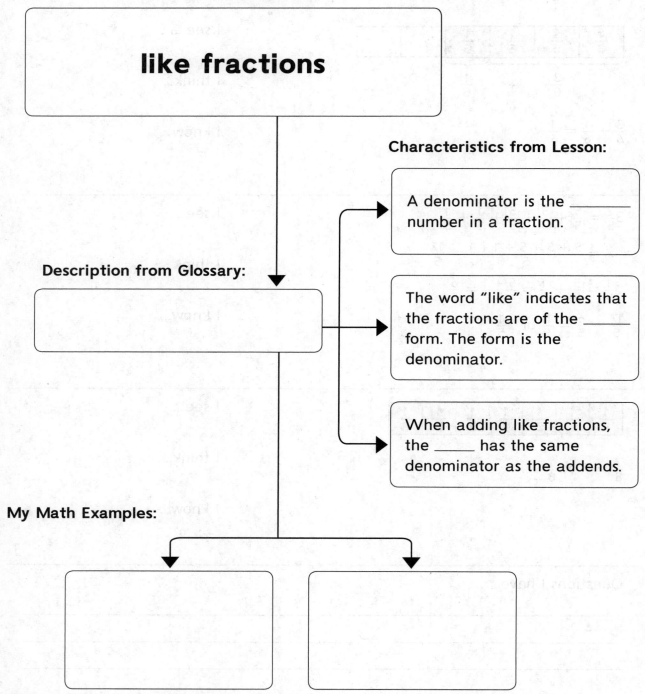

NAME ______________________ DATE ____________

Lesson 2 Vocabulary Chart

Add like Fractions

Use the three-column chart to organize the review vocabulary in this lesson. Write the word in Spanish. Then write the correct terms to complete each definition.

English	Spanish	Definition
denominator		The ______ number in a fraction. In $\frac{5}{6}$, ___ is the denominator.
numerator		The number ______ the bar in a fraction; the part of the fraction that tells how many of the equal parts are ______ ______.
simplest form		A fraction in which the numerator and denominator have ___ common factor greater than 1. $-$ is the simplest form of $\frac{6}{10}$.
greatest common factor		The ________ of the common factors of two or more numbers. The greatest common factor of 12, 18, and 30 is ___.
like fractions		Fractions that have the ______ denominator. $\frac{1}{5}$ and $\frac{2}{\ }$

NAME ______________________________ DATE ______________

Lesson 3 Guided Writing

Inquiry/Hands On: Use Models to Subtract Like Fractions

How do you subtract like fractions using models?

Use the exercises below to help you build on answering the Essential Question. Write the correct word or phrase on the lines provided.

1. Rewrite the question in your own words.

____

____

2. What key words do you see in the question?

____

3. A ________ fraction has a numerator of 1.

4. Is the fraction modeled below a unit fraction? ______

$\frac{1}{10}$

5. How many fraction tiles are used to model the fraction $\frac{9}{10}$? ______

6. How many fraction tiles are used to model the fraction $\frac{2}{10}$? ______

7. When you subtract 9 − 2, you take away ____ from ____.

8. When you subtract $\frac{9}{10} - \frac{2}{10}$, you take away ____ unit fractions.

9. Model $\frac{9}{10} - \frac{2}{10}$, find the difference. ______

10. How do you subtract like fractions using models?

____

____

____

____

NAME ______________________ DATE ____________

Lesson 4 Vocabulary Cognates

Subtract Like Fractions

Use the Glossary to define the math word in English and in Spanish in the word boxes. Write a sentence using your math word.

like fractions	**fracciones semejantes**
Definition	**Definición**
My math word sentence:	

simplest form	**mínima expresión**
Definition	**Definición**
My math word sentence:	

NAME ______________________ DATE ____________

Lesson 5 Problem-Solving Investigation

STRATEGY: Work Backward

Work backward to solve each problem.

1. **Chloe** did **some** of her homework **before** dinner.

   She did $\frac{2}{6}$ of her homework **after** dinner. She has $\frac{1}{6}$ of her homework **left**.

   What **fraction** of her homework did Chloe do **before** dinner?

   Write in simplest form.

Understand I know: I need to find:	**Solve**
Plan	**Check**

2. There were **12** goals scored **during** the game.

   Team A scored $\frac{8}{12}$ of the goals.

   Team B scored **2** goals during the **first half** of the game.

   What **fraction** of the goals did **Team B** score during the **second half** of the game?

   Write in simplest form.

Understand I know: I need to find:	**Solve**
Plan	**Check**

NAME ______________________ DATE ______________

Lesson 6 Note Taking

Add Mixed Numbers

Read the question. Write words you need help with and research each word. Use your lesson to write your Cornell notes. Write or draw math examples to explain your thinking. Share your examples with a classmate.

Building on the Essential Question How do you add mixed numbers?	**Notes:** A number that has a whole number part and a fraction part is a ________ ________. The mixed number $4\frac{2}{3}$ decomposed into a sum of whole numbers and unit fractions is equal to $\_\_ + \_\_ + \_\_ + \_\_ + \frac{1}{3} + \frac{1}{3}$. The mixed number $\_\_\frac{}{3}$ can be decomposed into $1 + 1 + \frac{1}{3}$. A fraction with a numerator that is greater than or equal to the denominator is an ________ ________.
Words I need help with:	The sum of the whole numbers, $1 + 1 + 1 + 1 + 1 + 1$ is ___. The sum of the unit fractions, $\frac{1}{3} + \frac{1}{3} + \frac{1}{3}$ is $\frac{}{3}$ I can use the equivalent fractions of one whole to write an improper fraction as a mixed number. $4\frac{2}{3} + 2\frac{1}{3} = \qquad = \_\_$
My Math Examples:	

NAME ____________________ DATE ____________

Lesson 7 Concept Web

Subtract Mixed Numbers

Use the concept web to write the equivalent improper fraction of each mixed number.

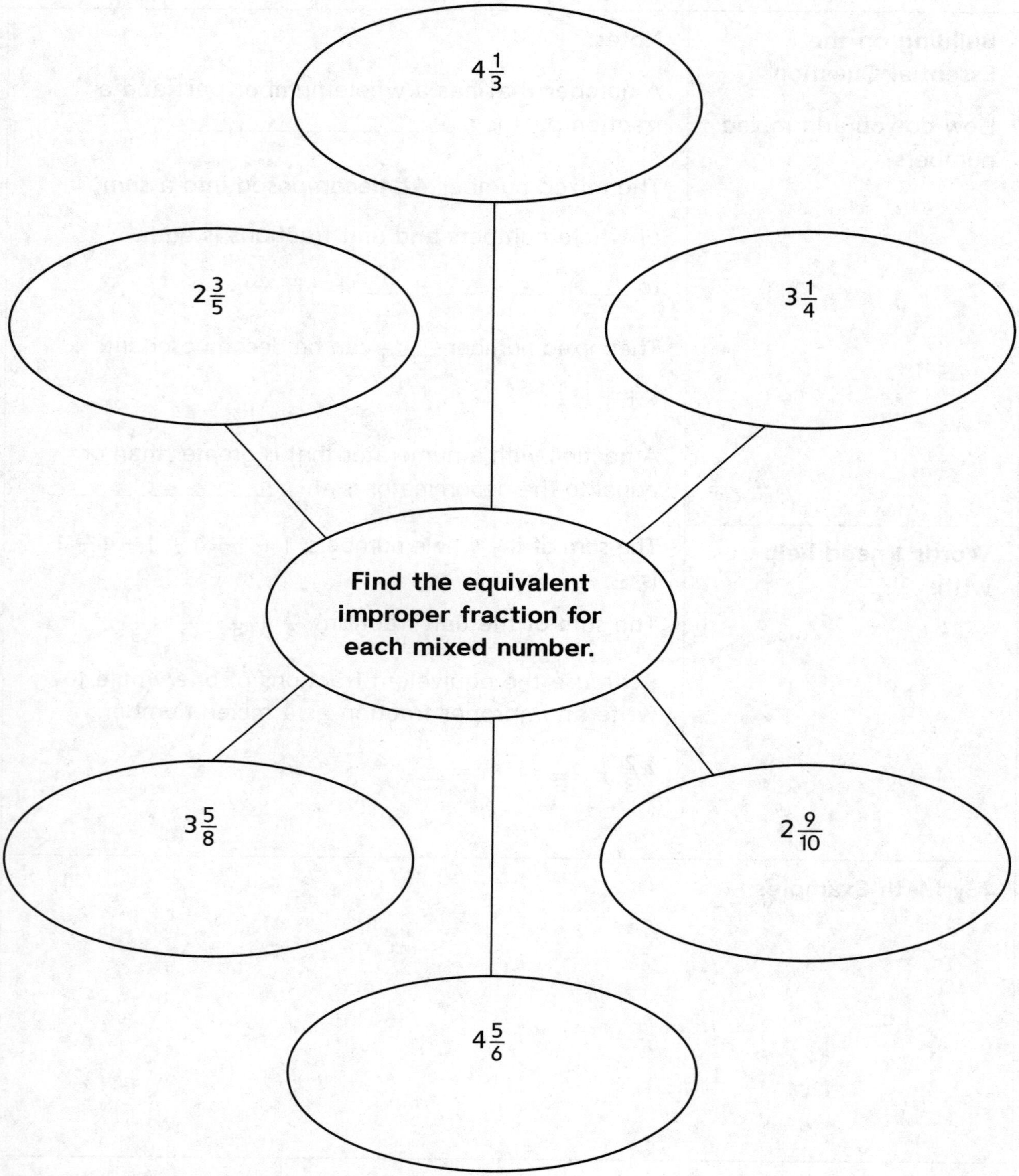

NAME ______________________________ DATE ______________

Lesson 8 Note Taking

Inquiry/Hands On: Model Fractions and Multiplication

Read the question. Write words you need help with and research each word. Use your lesson to write your Cornell notes. Write or draw math examples to explain your thinking. Share your examples with a classmate.

Building on the Essential Question

How do you model the multiplication of fractions?

Notes:

Find the sum. $2 + 2 + 2 =$ ____

Find the product. $3 \times 2 =$ ____

Multiplication can be thought of as repeated addition.

Two unit fraction tiles are used to model the fraction $\frac{\ }{8}$.

$\frac{1}{8}$	$\frac{1}{8}$

How many fraction tiles are used to model $\frac{2}{8} + \frac{2}{8} + \frac{2}{8}$? ______

How many fraction tiles are used to model $3 \times \frac{2}{8}$?

______

Find the sum. $\frac{2}{8} + \frac{2}{8} + \frac{2}{8} = \frac{\ }{8}$

Find the product. $3 \times \frac{2}{8} = \frac{\ }{8}$

Words I need help with:

My Math Examples:

NAME ______________________________ DATE ______________

Lesson 9 Guided Writing

Multiply Fractions by Whole Numbers

How do you multiply a fraction by a whole number?

Use the exercises below to help you build on answering the Essential Question. Write the correct word or phrase on the lines provided.

1. Rewrite the question in your own words.

____

____

2. What key words do you see in the question?

____

3. When you ____________ a number, you break it into different parts.

4. Decompose the fraction $\frac{3}{8}$ into the sum of a unit fraction.

____

5. Decompose the fraction $\frac{3}{8}$ into the product of a whole number and a unit fraction.

____

6. Use the Associative Property to rewrite the multiplication.

$7 \times \frac{3}{8} = 7 \times \left(3 \times \frac{1}{8}\right) = (\_\_ \times \_\_) \times \frac{}{8}$

7. Find the product of $7 \times \frac{3}{8}$

____

8. How do you multiply a fraction by a whole number?

____

____

____

NAME ______________________ DATE ______________

Chapter 10 Fractions and Decimals

Inquiry of the Essential Question:

How are fractions and decimals related?

Read the Essential Question. Describe your observations (I see...), inferences (I think...), and prior knowledge (I know...) of each math example. Write additional questions you have below. Then share your ideas and questions with a classmate.

$\frac{13}{100}$

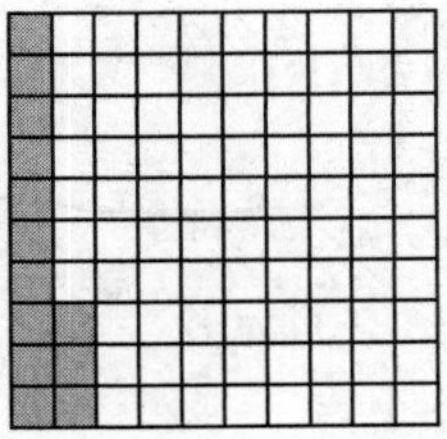

Ones	Tenths	Hundredths
0.	1	3

I see ...

I think...

I know...

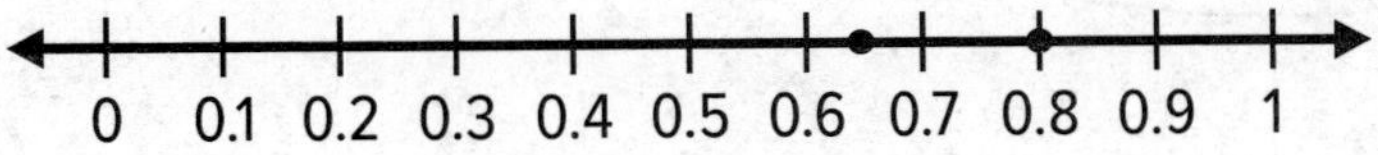

0.8 is to the right of 0.65. So, 0.8 > 0.65.

I see ...

I think...

I know...

Questions I have...

____

____

____

NAME ______________________ DATE ____________

Lesson 1 Concept Web

Inquiry/Hands On: Place Value Through Tenths and Hundredths

Use the concept web to write the place value of the 3 in each number.

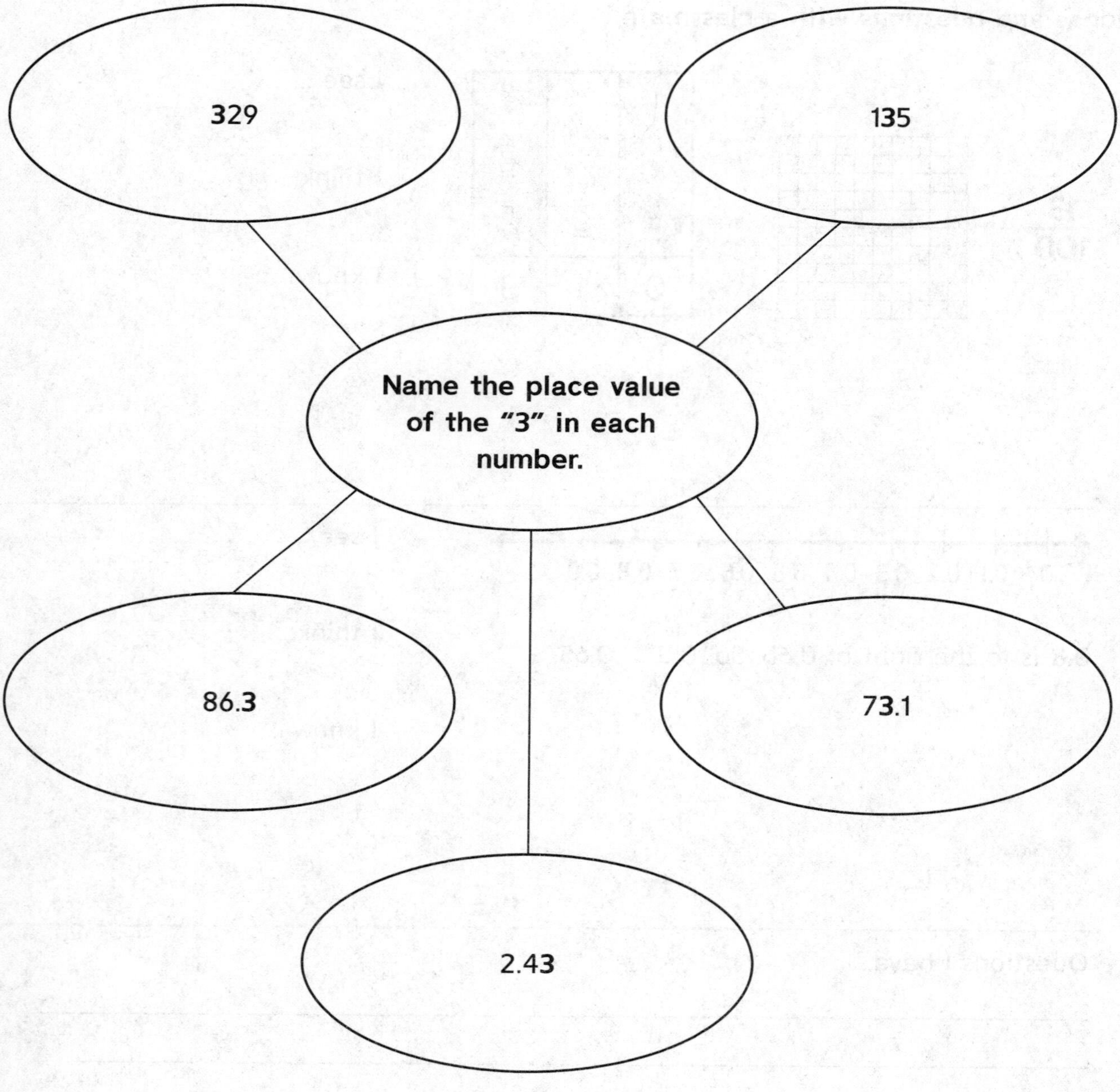

NAME ______________________________ DATE ______________

Lesson 2 Vocabulary Cognates

Tenths

Use the Glossary to define the math word in English and in Spanish in the word boxes. Write a sentence using your math word.

decimal	**decimal**
Definition	**Definición**
My math word sentence:	

tenth	**décima**
Definition	**Definición**
My math word sentence:	

NAME ______________________ DATE ____________

Lesson 3 Vocabulary Definition Map

Hundredths

Use the definition map to write a description and list characteristics about the vocabulary word or phrase. Write or draw math examples. Share your examples with a classmate.

My Math Vocabulary:

hundredth

Characteristics from Lesson:

In the number 3.45, the number ___ is in the hundredths place.

The hundredths place is located to the ______ of the tenths place.

One hundred pennies is __________ one dollar. One penny represents one ____________ of a dollar.

Description from Glossary:

My Math Examples:

NAME ________________________ DATE ____________

Lesson 4 Guided Writing

Inquiry/Hands On: Model Decimals and Fractions

How do you model decimals and fractions?

Use the exercises below to help you build on answering the Essential Question. Write the correct word or phrase on the lines provided.

1. Rewrite the question in your own words.

____

____

2. What key words do you see in the question?

____

3. The word form of the fraction — is *three tenths.*

4. Identify the decimal represented by the model shown. Write the decimal in the place-value chart.

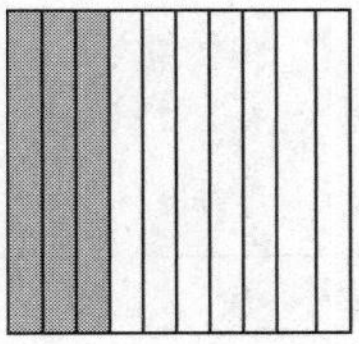

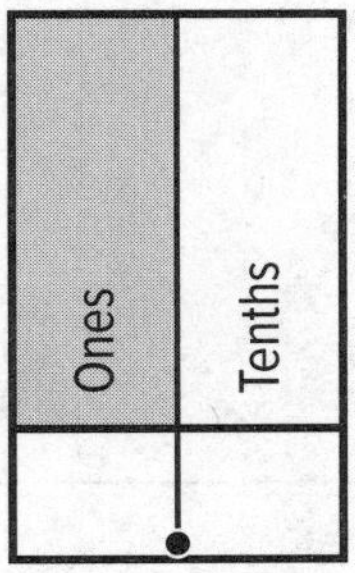

5. The word form of the fraction — is *thirty hundredths.*

6. Identify the decimal represented by the model shown. Write the decimal in the place-value chart.

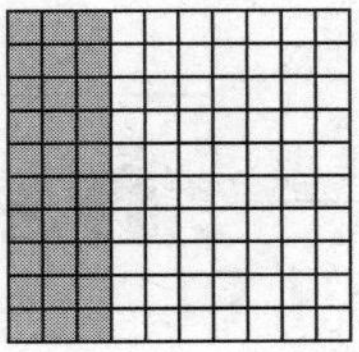

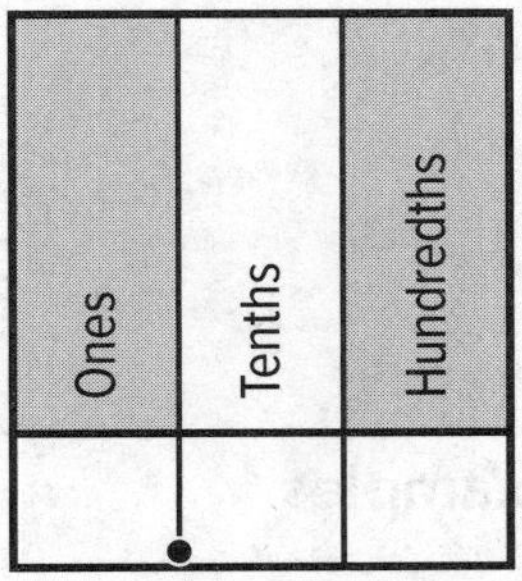

7. The decimals 0.3 and 0.30 are ____________. The fractions $\frac{3}{10}$ and $\frac{30}{100}$ are ____________.

8. How do you model decimals and fractions?

____

____

NAME ______________________________ DATE ______________

Lesson 5 Four-Square Vocabulary

Decimals and Fractions

Write the definition for each math word. Write what each word means in your own words. Draw or write examples that show each math word meaning. Then write your own sentences using the words.

Definition	My Own Words
decimal	
My Examples	My Sentence

Definition	My Own Words
fraction	
My Examples	My Sentence

NAME ______________________________ DATE ______________

Lesson 6 Note Taking

Use Place Value and Models to Add

Read the question. Write words you need help with and research each word. Use your lesson to write your Cornell notes. Write or draw math examples to explain your thinking. Share your examples with a classmate.

Building on the Essential Question

How can you use place value and models to add?

Words I need help with:

Notes:

The fraction $\frac{\quad}{10}$ is represented in the decimal model below.

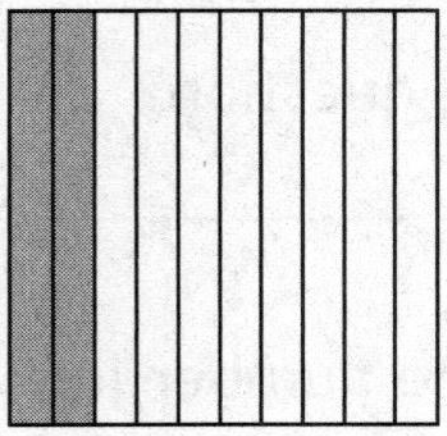

The fraction $\frac{\quad}{100}$ is represented in the decimal model below.

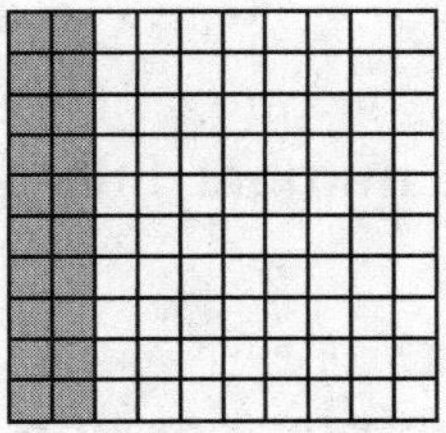

These two fractions are ____________.

Before adding $\frac{2}{10} + \frac{45}{100}$, write the fractions as like fractions.

Like fractions are fractions that have the same ______________.

$\frac{2}{10} + \frac{45}{100} = \frac{\quad}{100} + \frac{45}{100} = \frac{\quad}{100}$

The sum, written in decimal form, is _____.

My Math Examples:

NAME ______________________________ DATE ______________

Lesson 7 Guided Writing

Compare and Order Decimals

How do you compare and order decimals?

Use the exercises below to help you build on answering the Essential Question. Write the correct word or phrase on the lines provided.

1. Rewrite the question in your own words.

____

____

2. What key words do you see in the question?

____

3. When you move to the **right** on the number line below, the whole numbers ________ in value.

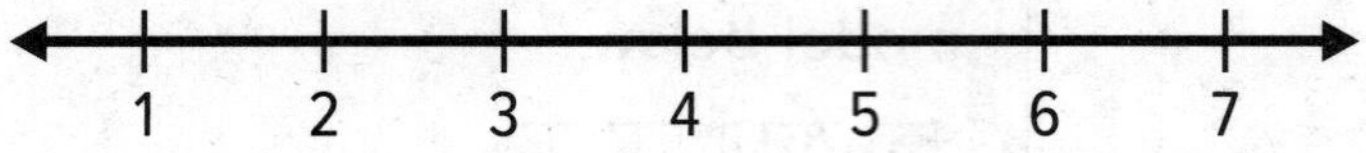

4. When you move to the **left** on the number line above, the whole numbers ________ in value.

5. When you move to the **right** on the number line below, the decimals ________ in value.

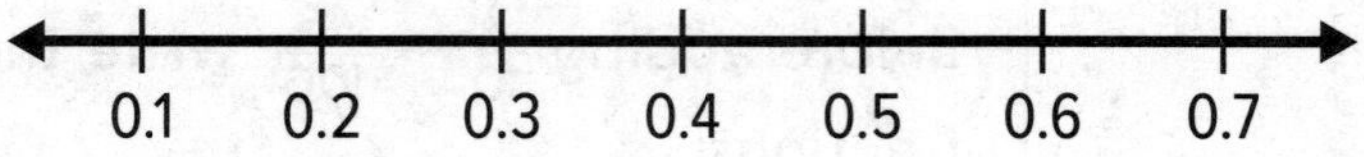

6. When you move to the **left** on the number line above, the decimals ________ in value.

7. How do you compare and order decimals?

____

____

NAME _______________ DATE _______________

Lesson 8 Problem-Solving Investigation

STRATEGY: Extra or Missing Information

Determine if there is extra or missing information to solve each problem. Then solve if possible.

1. There are **100 movies** at the store.
$\frac{30}{100}$ are **action** movies, $\frac{50}{100}$ are **comedies**, and $\frac{20}{100}$ are **adventure** movies.
What part of the movies are **action** or **comedies**?

Understand I know: I need to find:	**Solve**
Plan The extra or missing information is:	**Check**

2. In a basketball game, the **red team** scored $\frac{3}{10}$ of the baskets during the **first half** and $\frac{4}{10}$ of the baskets during the **second half**.
The blue team had 10 players.
How many baskets did the **red team** score during the **first half and second half** of the game?

Understand I know: I need to find:	**Solve**
Plan The extra or missing information is:	**Check**

NAME ________________________________ DATE ______________

Chapter 11 Customary Measurement

Inquiry of the Essential Question:

Why do we convert measurements?

Read the Essential Question. Describe your observations (I see...), inferences (I think...), and prior knowledge (I know...) of each math example. Write additional questions you have below. Then share your ideas and questions with a classmate.

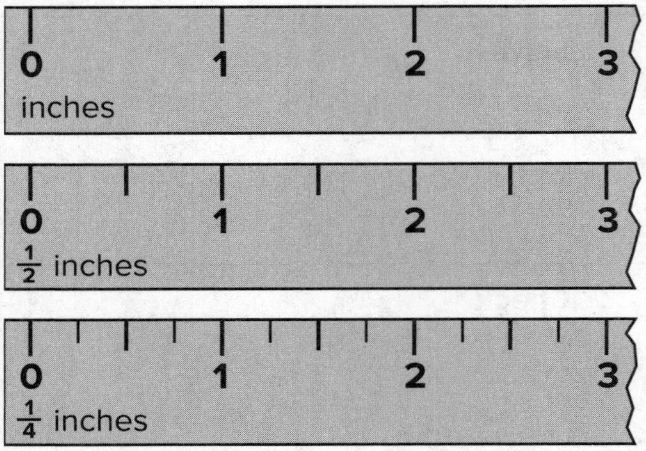

I see ...

I think...

I know...

seconds (s)	minutes (min)	(s, min)
60	1	(60, 1)
120	2	(120, 2)
180	3	(180, 3)
240	4	(240, 4)

I see ...

I think...

I know...

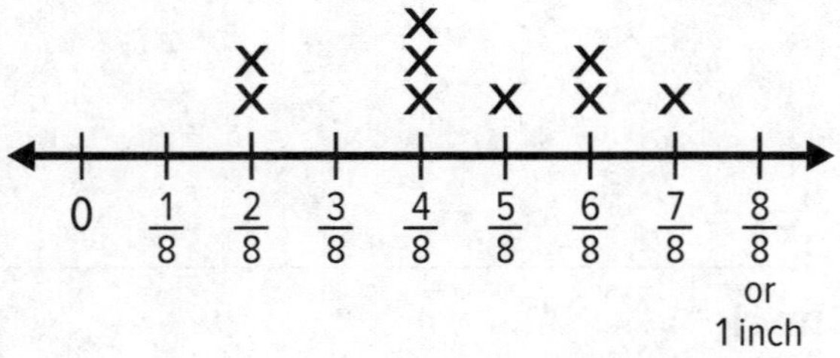

I see ...

I think...

I know...

Questions I have...

____

____

____

NAME ________________________ DATE ____________

Lesson 1 Multiple Meaning Word

Customary Units of Length

Complete the four-square chart to review the multiple meaning words *foot* and *yard*.

Everyday Use	**Math Use in a Sentence**
Math Use	**Example From This Lesson**

foot (ft)

Everyday Use	**Math Use in a Sentence**
Math Use	**Example From This Lesson**

yard (yd)

Write the correct numbers on each line to complete the sentence.

Since 1 yard = ___ feet and 1 foot = ___ inches, then you know 3 feet = ___ inches and 1 yard = ___ inches.

NAME _______________ DATE _______________

Lesson 2 Vocabulary Cognates

Convert Customary Units of Length

Use the Glossary to define the math word in English and in Spanish in the word boxes. Write a sentence using your math word.

convert	**convertir**
Definition	**Definición**
My math word sentence:	

mile (mi)	**milla (mi)**
Definition	**Definición**
My math word sentence:	

NAME ______________________________ DATE ____________

Lesson 3 Vocabulary Definition Map

Customary Units of Capacity

Use the definition map to write a description and list characteristics about the vocabulary word or phrase. Write or draw math examples. Share your examples with a classmate.

My Math Vocabulary:

capacity

Characteristics from Lesson:

The smallest customary unit of capacity is ____________.

A cup is ____ than a pint.

A quart is ________ than a pint.

The largest customary unit of capacity is a ________.

Description from Glossary:

My Math Examples:

NAME ______________________ DATE ______________

Lesson 4 Vocabulary Chart

Convert Customary Units of Capacity

Use the three-column chart to organize the review vocabulary in this lesson. Write the word in Spanish. Then write the correct terms to complete each definition.

English	Spanish	Definition
capacity		The amount of ______ a container can hold.
convert		To _______ one unit to another.
is equal to (=)		Having the ______ value. The = sign is used to show two numbers or expressions are ______.
is greater than (>)		An __________ relationship showing that the number on the ____ side of the symbol is ________ than the number on the right side.
is less than (<)		An __________ relationship showing that the number on the ____ side of the symbol is ____ than the number on the right side.

NAME ______________________________ DATE ______________

Lesson 5 Concept Web

Customary Units of Weight

Use the concept web to write the best unit of weight for each object. The first one is done for you.

Word Bank		
ounces	pounds	tons
(A slice of bread weighs about 1 oz.)	(A football weighs about 1 lb.)	(A buffalo weighs about 1 T.)

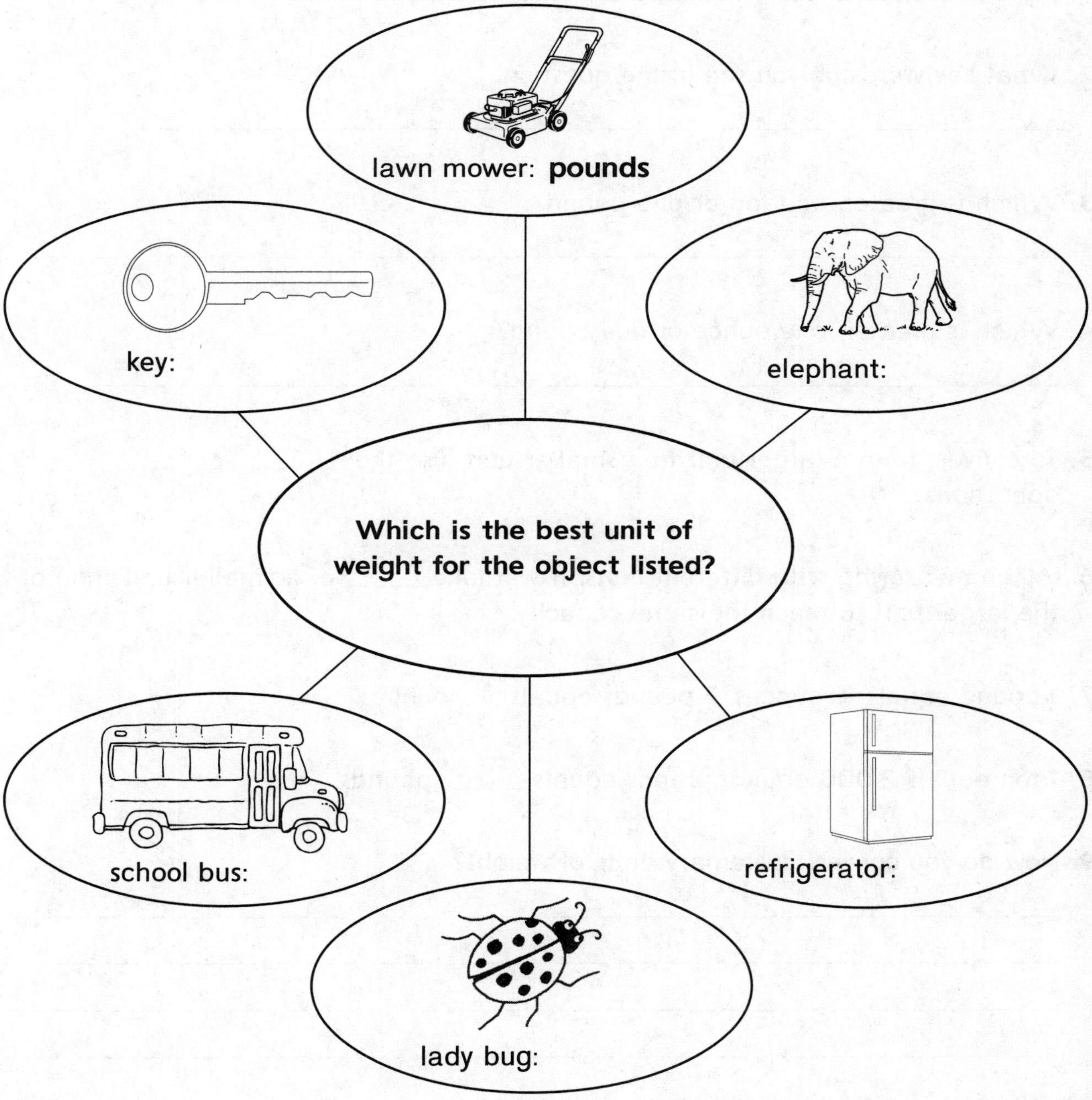

NAME ______________________ DATE ______________

Lesson 6 Guided Writing

Convert Customary Units of Weight

How do you convert customary units of weight?

Use the exercises below to help you build on answering the Essential Question. Write the correct word or phrase on the lines provided.

1. Rewrite the question in your own words.

____

____

2. What key words do you see in the question?

____

3. Which is greater, one ton or one pound?

____

4. Which is greater, one ounce or one pound?

____

5. To convert from a larger unit to a smaller unit, use the ______________ operation.

6. When measuring with different units, it will take _____ of a smaller unit than of the larger unit to reach the same capacity.

7. 1 pound equals 16 ounces. 2 pounds equals ___ ounces.

8. 1 ton equals 2,000 pounds. 3 tons equals _____ pounds.

9. How do you convert customary units of weight?

____

____

____

____

NAME ______________________ DATE ______________

Lesson 7 Note Taking

Convert Units of Time

Read the question. Write words you need help with and research each word. Use your lesson to write your Cornell notes. Write or draw math examples to explain your thinking. Share your examples with a classmate.

Building on the Essential Question How do you convert units of time?	**Notes:** Years, months, weeks, days, hours, minutes, and seconds are all different units of __________. • One hour is __________ than one minute. • One second is __________ than one minute. • One day is __________ than one hour. • One week is __________ than one day. • One month is __________ than one year. To convert from a larger unit to a smaller unit, use the __________ operation.
Words I need help with:	1 day equals ________ hours. $(1 \times 2) = 2$ 2 days equals ________ hours. 1 week equals ________ days. $(1 \times 3) = 3$ 3 weeks equals ________ days. To find the number of minutes in 3 hours, multiply the number of minutes in 1 hour by the number ________. 1 hour equals ________ minutes. 3 hours equals ________ minutes.
My Math Examples:	

NAME ______________________________ DATE ______________

Lesson 8 Vocabulary Definition Map

Display Measurement Data in a Line Plot

Use the definition map to write a description and list characteristics about the vocabulary word or phrase. Write or draw math examples. Share your examples with a classmate.

My Math Vocabulary:

line plot

Description from Glossary:

Characteristics from Lesson:

A ________ ________ is a line with numbers on it, in order, at regular intervals.

________ is numbers or symbols, sometimes collected from a survey or experiment, to show information.

You can represent ______________ data for fractions of a unit in a line plot.

My Math Examples:

NAME _______________ DATE _______________

Lesson 9 Note Taking

Solve Measurement Problems

Read the question. Write words you need help with and research each word. Use your lesson to write your Cornell notes. Write or draw math examples to explain your thinking. Share your examples with a classmate.

Building on the Essential Question How do you solve measurement problems in the customary system? **Words I need help with:**	**Notes:** When solving measurement problems, all values need to be the __________ unit __________ performing operations. There are ______ minutes in 1 hour. 1 hour − 15 minutes = ? ______ minutes − 15 minutes = ______ minutes There are ______ ounces in 1 pound. 3 pounds + 10 ounces = ? (______ × 3) ounces + 10 ounces = ? ______ ounces + 10 ounces = ______ ounces When a problem asks for the **difference** in units, use the operation of __________ . When a problem asks for the **total** number of units, use the operation of __________.
My Math Examples:	

NAME ______________________________ DATE ______________

Lesson 10 Problem-Solving Investigation

STRATEGY: Guess, Check, and Revise

Guess, check and revise to solve each problem.

1. **Max** was on vacation **twice** as long as **Jared** and **half** as long as **Wesley**. The **boys** were on vacation a **total** of **3 weeks**. How many **days** was **each** boy on vacation?

Understand

I know:

I need to find:

Solve

Name	Days
Max	
Jared	
Wesley	
Total	

Plan

Find how many days = 1 week.

Check

2. Anu drinks **2 cups** of water each day. Jan drinks **twice** as much water as **Anu**. How many **fluid ounces** does **Jan** drink?

Understand

I know:

I need to find:

Solve

Plan

Find how many fluid ounces = 1 cup.

Check

NAME ________________________________ DATE ______________

Chapter 12 Metric Measurement

Inquiry of the Essential Question:

How can conversion of measurements help me solve real-world problems?

Read the Essential Question. Describe your observations (I see...), inferences (I think...), and prior knowledge (I know...) of each math example. Write additional questions you have below. Then share your ideas and questions with a classmate.

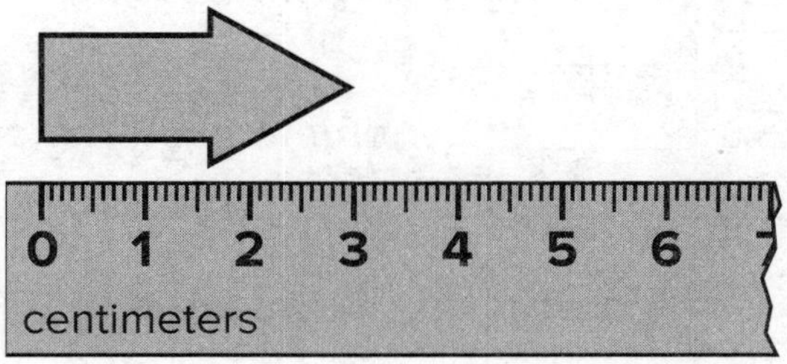

The arrow is about ________ centimeters long.

I see ...

I think...

I know...

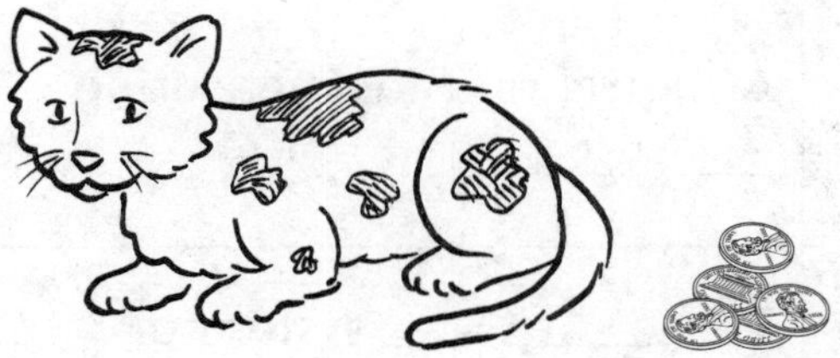

Mass of cat: 5 grams or 5 kilograms?

THINK: 5 grams would have the same mass as about 5 pennies.

I see ...

I think...

I know...

Conversion Table

Kilograms (kg)	grams (g)	(kg, g)
12	12,000	(12,12,000)
14	14,000	(14,14,000)
16	16,000	(16,16,000)
18	18,000	(18,18,000)

I see ...

I think...

I know...

Questions I have...

____

____

____

NAME ______________________ DATE ______________

Lesson 1 Vocabulary Chart

Metric Units of Length

Use the three-column chart to organize the vocabulary in this lesson. Write the word in Spanish. Then write the correct terms to complete each definition.

English	Spanish	Definition
centimeter (cm)		A metric unit for measuring ________. ________ centimeters = 1 meter
kilometer (km)		A metric unit for measuring ________. 1 kilometer = ________ meters
meter (m)		A metric unit for measuring ________.
metric system (SI)		The __________ system of measurement. Includes units *such* as *meter*, ________, and *liter*.
millimeter (mm)		A metric unit for measuring ________. ________ millimeters = 1 meter

NAME ______________________ DATE ____________

Lesson 2 Vocabulary Cognates

Metric Units of Capacity

Use the Glossary to define the math word in English and in Spanish in the word boxes. Write a sentence using your math word.

liter (L)	**litro (L)**
Definition	**Definición**
My math word sentence:	

milliliter (mL)	**mililitro (mL)**
Definition	**Definición**
My math word sentence:	

NAME ____________________ DATE ____________

Lesson 3 Concept Web

Metric Units of Mass

Use the concept web to write the best unit of mass to use for each object. The first one is done for you.

Word Bank	
grams	kilograms
(A paperclip weighs about 1 g.)	(A loaf of bread weighs about 1 kg.)

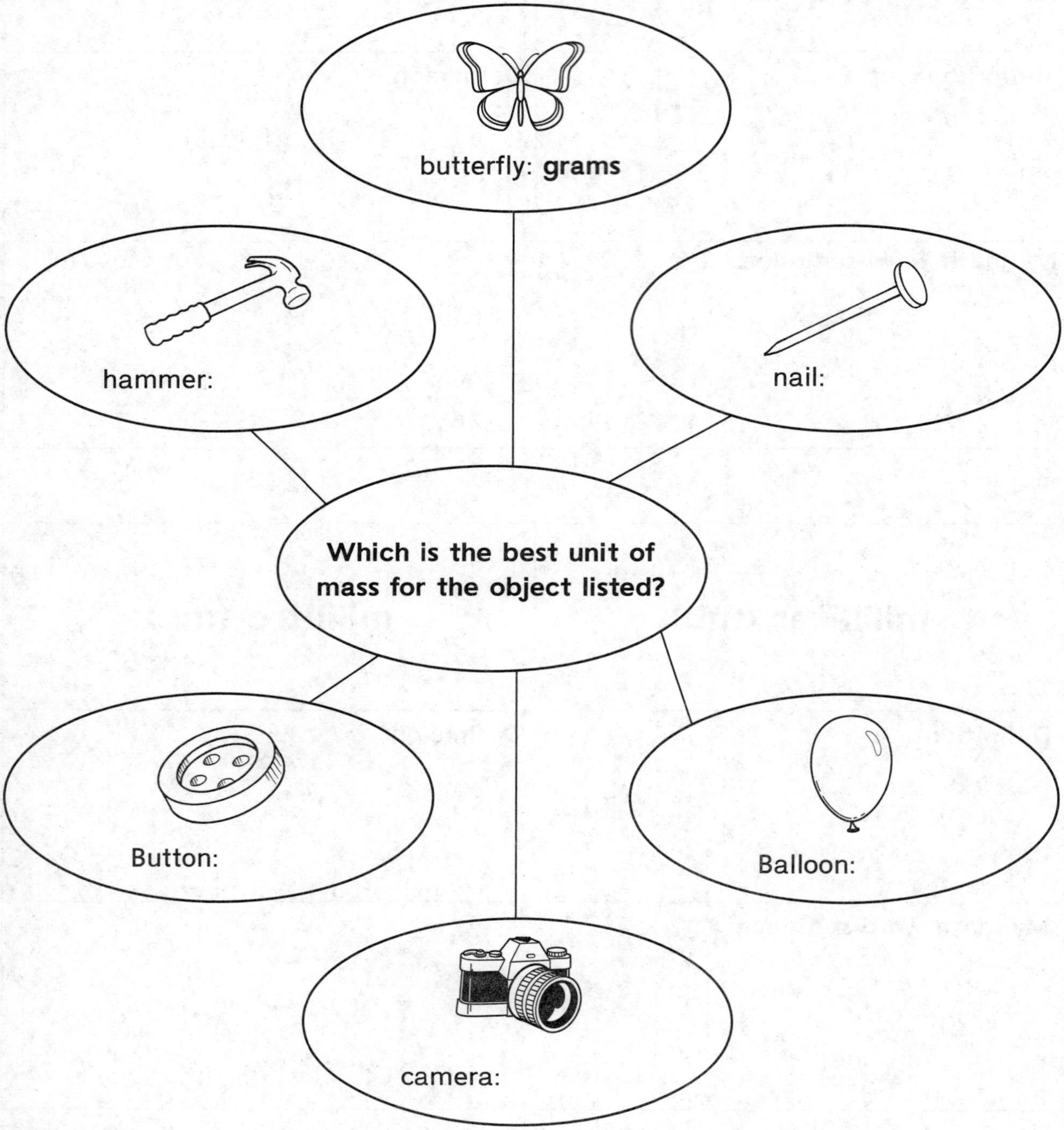

NAME ______________________________ DATE ______________

Lesson 4 Problem-Solving Investigation

STRATEGY: Make an Organized List

Make an organized list to solve each problem.

1. **Brianna** has 0.16 of a dollar.
   How many different combinations of coins could **she** (Brianna) have?

Understand	Solve
I know: I need to find:	
Plan 1 penny = 0.01 of a dollar 1 nickel = ______ of a dollar 1 dime = ______ of a dollar 1 quarter = ______ of a dollar	**Check**

2. There were three races at the track meet.
   The distances were **100** meters long, **800** meters long and **3,200** meters long.
   Suppose **Lucy** ran two of the races.
   What are the possible total distances that **she** (Lucy) ran?

Understand

I know:

I need to find:

Solve

Lucy's 1st Race	Lucy's 2nd Race	Total Distance

Plan

Check

NAME ______________________________ DATE ______________

Lesson 5 Guided Writing

Convert Metric Units

How do you convert metric units?

Use the exercises below to help you build on answering the Essential Question. Write the correct word or phrase on the lines provided.

1. Rewrite the question in your own words.

____

____

2. What key words do you see in the question?

____

3. Which is a greater unit of measure, one meter or one centimeter?

____

4. Which is a greater unit of measure, one gram or one kilogram?

____

5. To convert from a larger unit to a smaller unit, use the ____________ operation.

6. ________ equivalencies are all multiples of 10, 100, and 1,000.

7. 1 kilogram equals 1,000 grams. 2 kilograms equals ________ grams.

8. 1 meter equals 100 centimeters. 3 meters equals ________ centimeters.

9. How do you convert metric units?

____

____

NAME ________________________________ DATE ______________

Lesson 6 Note Taking

Solve Measurement Problems

Read the question. Write words you need help with and research each word. Use your lesson to write your Cornell notes. Write or draw math examples to explain your thinking. Share your examples with a classmate.

Building on the Essential Question How do you solve measurement problems in the metric system?	**Notes:** Metric units for measurement of length are ____________ (cm), ________ (m), and ____________ (km). Metric units for measurement of mass are ________ (g) and ____________ (kg). Metric units for measurement of capacity are ________ (L) and ____________ (mL). When solving measurement problems, all values need to be the ________ unit ________ performing operations. There are ________ milliliters in 1 liter.
Words I need help with:	1 liter − 150 milliliters = ? _______ milliliters − 150 milliliters = _______ milliliters There are ____ millimeters in 1 centimeter. 5 centimeters + 8 millimeters = ? (____ × 5) centimeters + 8 millimeters = ? ____ millimeters + 8 millimeters = ____ millimeters When a problem asks for the **difference** in units, use the operation of ____________. When a problem asks for the **total** number of units, use the operation of __________.
My Math Examples:	

NAME ______________________ DATE ____________

Chapter 13 Perimeter and Area

Inquiry of the Essential Question:

Why is it important to measure perimeter and area?

Read the Essential Question. Describe your observations (I see...), inferences (I think...), and prior knowledge (I know...) of each math example. Write additional questions you have below. Then share your ideas and questions with a classmate.

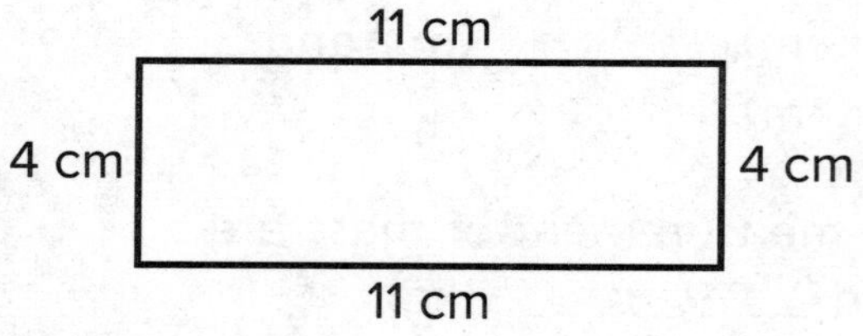

P = Perimeter

P = 11 cm + 4 cm + 11 cm + 4 cm = 30 cm

I see ...

I think...

I know...

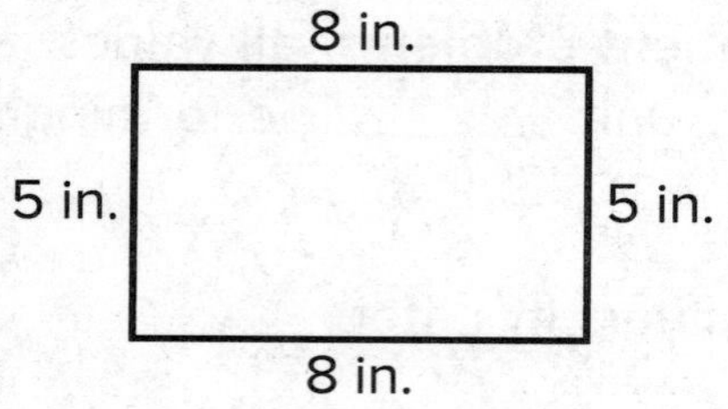

$P = (2 \times \ell) + (2 \times w)$

$P = (2 \times 8) + (2 \times 5)$

P = 16 + 10 or 26 in.

I see ...

I think...

I know...

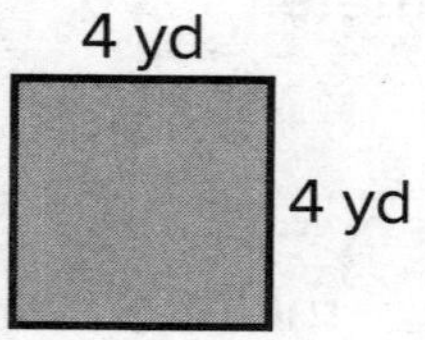

A = Area

$A = s \times s$

A = 4 yd × 4 yd

A = 16 sq yd

I see

I think...

I know...

Questions I have...

____

____

____

NAME ______________________________ DATE ______________

Lesson 1 Vocabulary Definition Map

Measure Perimeter

Use the definition map to write a description and list characteristics about the vocabulary word or phrase. Write or draw math examples. Share your examples with a classmate.

My Math Vocabulary:

perimeter

Characteristics from Lesson:

A ________ figure is a shape without any openings.

Description from Glossary:

The perimeter is the ________ of the measurements of each side.

Perimeter formulas for a rectangle and a square are:

__________: $P = 2\ell + 2w$

__________: $P = 4s$

My Math Examples:

NAME ____________________ DATE ____________

Lesson 2 Problem-Solving Investigation

STRATEGY: Solve a Simpler Problem

Solve each problem by solving a simpler problem.

1. Clarissa has four **pictures** that are each the **size** of the one **shown**. What will be the perimeter of the rectangle formed if the four pictures are laid end to end as show?

Understand	**Solve**
Plan	**Check**

2. Mr. and Mrs. Lopez are putting **square tiles** on the floor in their bathroom. **They** (Mr. and Mrs. Lopez) can fit **6 rows** of **4 tiles** in the bathroom. How many **tiles** do they **need** to buy? If **each** tile costs **$5**, what is the **total cost**?

Understand	**Solve**
Plan 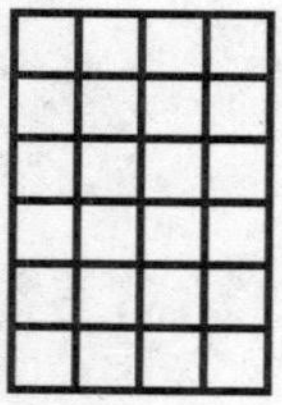	**Check**

NAME ______________________________ DATE ______________

Lesson 3 Guided Writing

Inquiry/Hands On: Model Area

How do you model area?

Use the exercises below to help you build on answering the Essential Question. Write the correct word or phrase on the lines provided.

1. Rewrite the question in your own words.

____

____

2. What key words do you see in the question?

____

3. A square is a rectangle that has _____ sides of the same length. The length and width are _______ in a square.

4. The side length of the square below is ___ unit(s).

5. The side length of the square below is ___ unit(s).

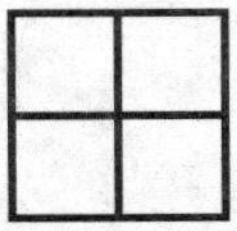

6. ______ is the number of square units needed to cover the inside of a region or plane figure without any overlap.

7. A unit square is different from a square unit. A _____ ________ is a square with a side length of one unit. ________ _____ is a unit for measuring area.

8. The area of the square below is ___ square unit(s).

9. The area of the square below is ___ square unit(s).

10. How do you model area?

____

____

____

NAME ________________________________ DATE ________________

Lesson 4 Multiple Meaning Word

Measure Area

Complete the four-square chart to review the multiple meaning word or phrase.

Everyday Use	**Math Use in a Sentence**
Math Use	**Example From This Lesson**

area

Write the correct term on each line to complete the sentence.

To find the ______ of a figure you can count the number of unit squares or multiple the _______ by ______.

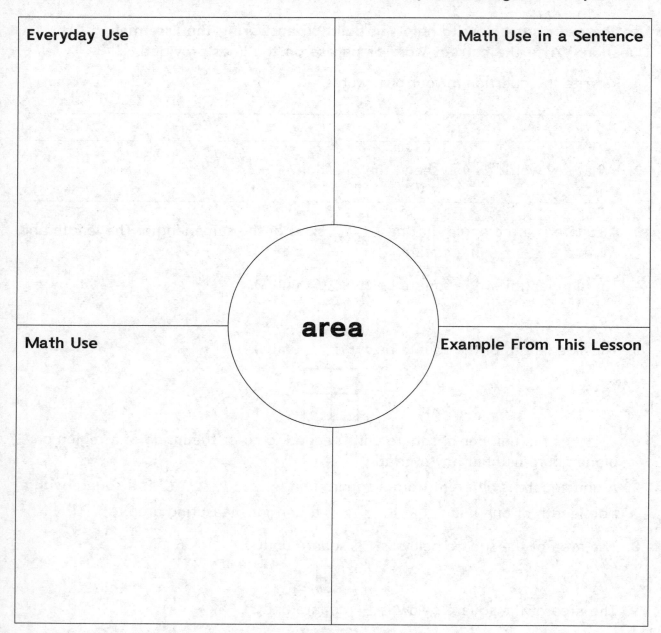

NAME ____________________ DATE ____________

Lesson 5 Note Taking

Relate Area and Perimeter

Read the question. Write words you need help with and research each word. Use your lesson to write your Cornell notes. Write or draw math examples to explain your thinking. Share your examples with a classmate.

Building on the Essential Question

How are area and perimeter related?

Notes:

Area is the number of ________ ______ needed to cover the ______ of a region or plane figure without any _________.

Perimeter is the _________ _________ a shape or region.

Two figures ____ have the same ___________ and different areas.

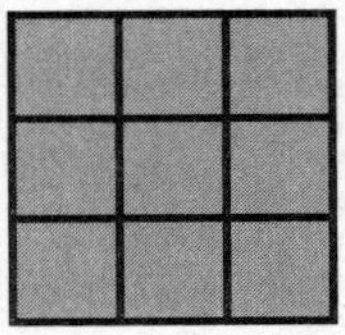

Perimeter: ___ units

Area: ___ square units

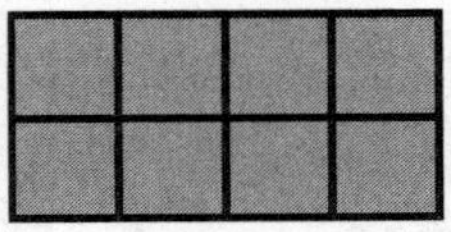

Perimeter: ___ units

Area: ___ square units

Two figures ____ have the same _____ and different perimeters.

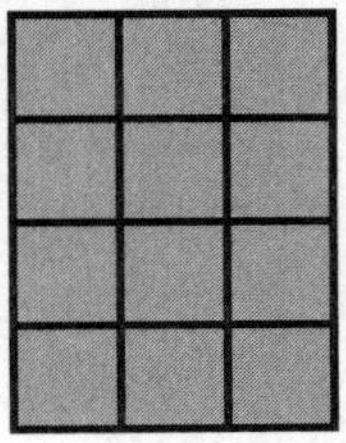

Perimeter: ___ units

Area: ___ square units

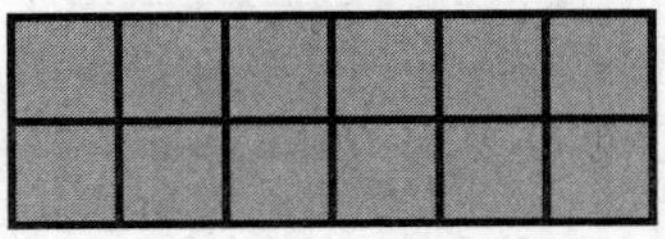

Perimeter: ___ units

Area: ___ square units

Words I need help with:

My Math Examples:

NAME ______________________ DATE ______________

Chapter 14 Geometry

Inquiry of the Essential Question:

How are different ideas about geometry connected?

Read the Essential Question. Describe your observations (I see...), inferences (I think...), and prior knowledge (I know...) of each math example. Write additional questions you have below. Then share your ideas and questions with a classmate.

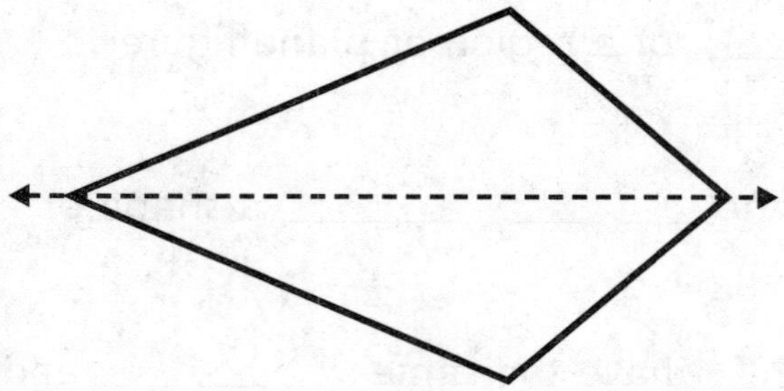

I see ...

I think...

I know...

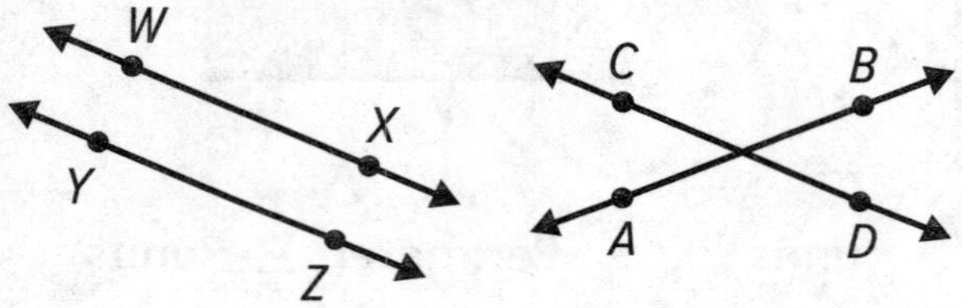

I see ...

I think...

I know...

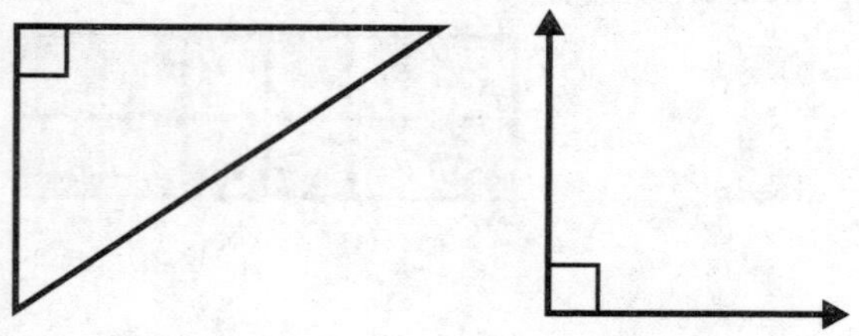

I see ...

I think...

I know...

Questions I have...

____

____

____

NAME ______________________________ DATE ______________

Lesson 1 Vocabulary Chart

Draw Points, Lines, and Rays

Use the three-column chart to organize the vocabulary in this lesson. Write the word in Spanish. Then write the correct terms to complete each definition.

English	Spanish	Definition
line		A ________ set of points that extend in ________ directions without ending.
line segment		A part of a ______ between two __________. The length of the line segment can be __________.
endpoint		The _____ at either end of a ____ ________ or the _____ at the beginning of a ___.
point		An _____ location in space that is represented by a _____.
ray		A part of a ______ that has one __________ and extends in one direction without ending.

NAME ________________________________ DATE ________________

Lesson 2 Vocabulary Definition Map

Draw Parallel and Perpendicular Lines

Use the definition map to write a description and list characteristics about the vocabulary word or phrase. Write or draw math examples. Share your examples with a classmate.

My Math Vocabulary:

line

Characteristics from Lesson:

__________ ______ are lines that meet or cross at a ______.

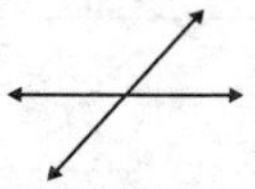

__________ ______ are lines that meet or cross each other to form ______ angles.

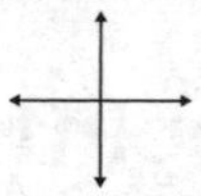

______ ____ are lines that are the same distance apart. ______ _____ do not meet.

Description from Glossary:

My Math Examples:

NAME ______________________ DATE ____________

Lesson 3 Guided Writing

Inquiry/Hands On: Model Angles

How do you model angles?

Use the exercises below to help you build on answering the Essential Question. Write the correct word or phrase on the lines provided.

1. Rewrite the question in your own words.

____

____

2. What key words do you see in the question?

____

3. Identify if each situation below models a $\frac{1}{4}$ turn, a $\frac{1}{2}$ turn, or a full turn.

$\frac{1}{4}$ turn

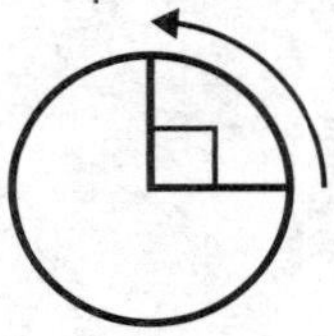

$\frac{1}{2}$ turn

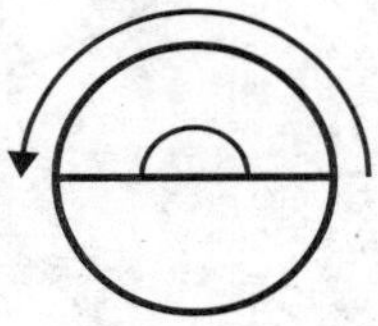

1 full turn

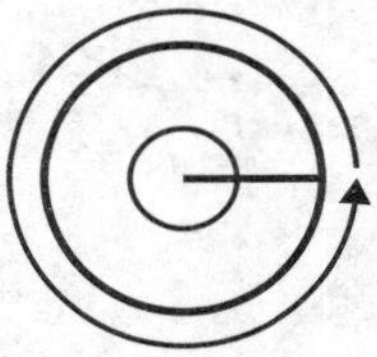

a. Extend both arms together out in front of you and then move one arm straight out to the side. ________

b. A clock's minute hand was on the 12, then moved to the 6. ________

4. A ______ is a part of a line that has one endpoint and extends in one direction without ending.

5. An ______ is a figure that is formed by two rays with the same endpoint.

6. Identify if each angle models as a $\frac{1}{4}$ turn, a $\frac{1}{2}$ turn, or a full turn between the two rays.

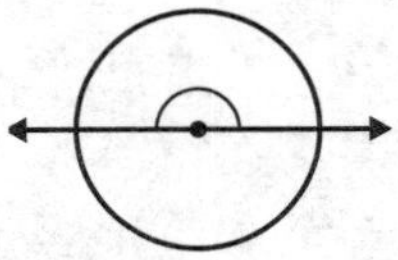

a. ________

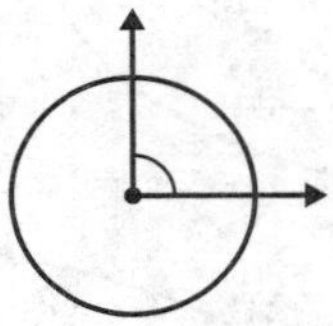

b. ________

7. How do you model angles?

____

____

NAME ______________________________ DATE ______________

Lesson 4 Concept Web

Classify Angles

Use the concept web to write if the angle is classified as *acute, obtuse,* or *right* based on the drawing or measurement given.

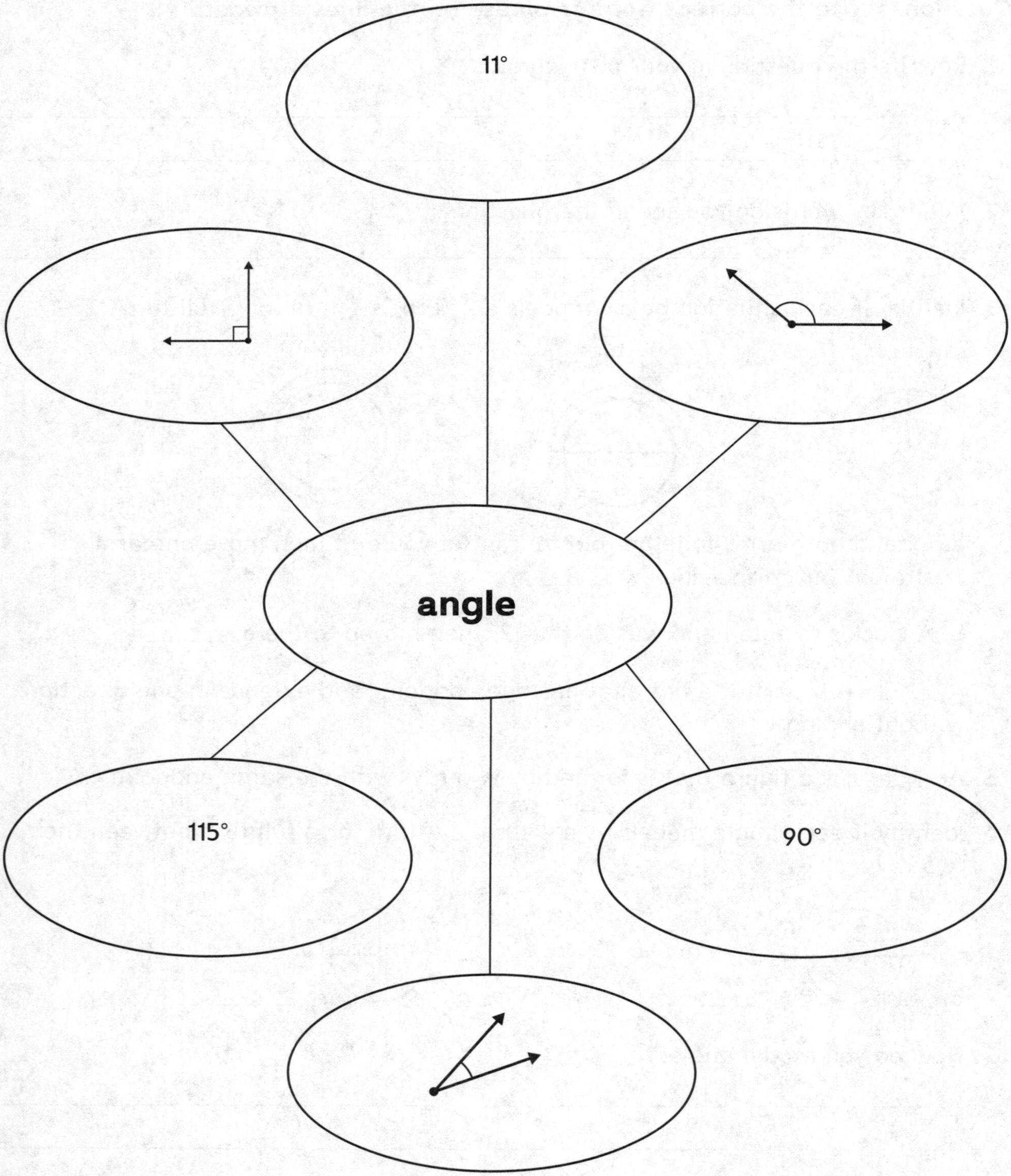

NAME ______________________________ DATE ______________

Lesson 5 Multiple Meaning Word

Measure Angles

Complete the four-square chart to review the multiple meaning word.

Everyday Use	**Math Use in a Sentence**
Math Use	**Example From This Lesson**

angle

Write the correct term on each line to complete the sentence.

Three types of angles are ______, ______, and ______.

NAME ______________________ DATE ______________

Lesson 6 Vocabulary Cognates

Draw Angles

Use the Glossary to define the math word in English and in Spanish in the word boxes. Write a sentence using your math word.

angle	**ángulo**
Definition	**Definición**
My math word sentence:	

ray	**semirrecta**
Definition	**Definición**
My math word sentence:	

NAME ____________________ DATE ____________

Lesson 7 Note Taking

Solve Problems with Angles

Read the question. Write words you need help with and research each word. Use your lesson to write your Cornell notes. Write or draw math examples to explain your thinking. Share your examples with a classmate.

Building on the Essential Question

How can you solve problems with angles?

Notes:

To decompose a number means to ______ a number into ________ parts.

____ = 50 + 40

An angle can also be decomposed into ________ parts.

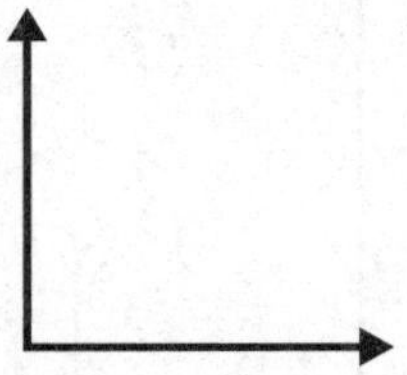

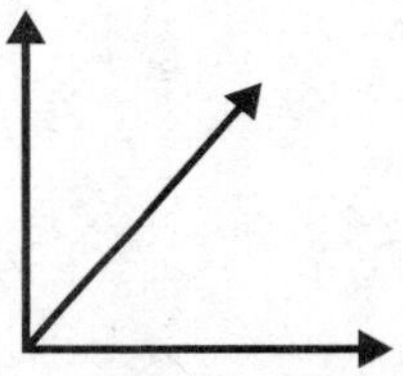

____ = 50° + 40°

Words I need help with:

A variable is a letter or symbol used to represent an ________ quantity.

When solving a problem to find an angle measure, you can use a ________ to represent the unknown angle measure.

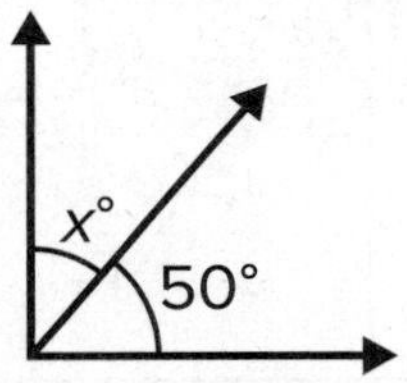

$90° - 50° = x° =$ ____

My Math Examples:

NAME _______________ DATE _______________

Lesson 8 Vocabulary Definition Map

Triangles

Use the definition map to write a description and list characteristics about the vocabulary word or phrase. Write or draw math examples. Share your examples with a classmate.

My Math Vocabulary:

triangle

Description from Glossary:

Characteristics from Lesson:

An ______ triangle is a triangle with one obtuse angle.

An ______ triangle is a triangle with all three angles less than 90°.

A ______ triangle is a triangle with one right angle.

My Math Examples:

NAME ______________________ DATE ______________

Lesson 9 Vocabulary Chart

Quadrilaterals

Use the three-column chart to organize the vocabulary in this lesson. Write the word in Spanish. Then write the correct terms to complete each definition.

English	Spanish	Definition
parallelogram		A quadrilateral in which each pair of opposite sides are ________ and equal in length.
rectangle		A quadrilateral with _____ right angles; opposite sides are equal and ________.
rhombus		A parallelogram with _____ congruent sides.
trapezoid		A quadrilateral with exactly _____ pair of ________ sides.
square		A rectangle with _____ congruent sides.

NAME ______________________ DATE ______________

Lesson 10 Note Taking

Draw Lines of Symmetry

Read the question. Write words you need help with and research each word. Use your lesson to write your Cornell notes. Write or draw math examples to explain your thinking. Share your examples with a classmate.

Building on the Essential Question

How do you draw lines of symmetry?

Notes:

A figure has ______________ if it can be folded so that the two parts of the figure match, or are congruent.

A _____ **of** __________ is a line on which a figure can be folded so that its two halves match exactly.

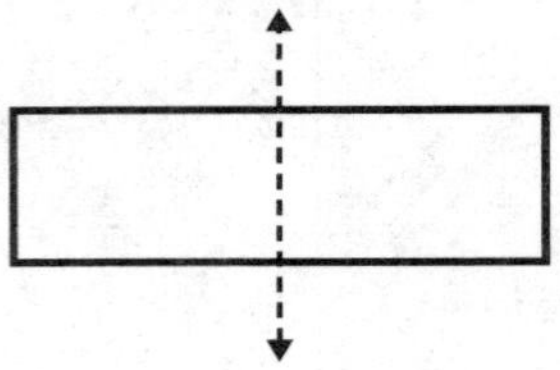

Figures **can** have more than one ______ of ___________.

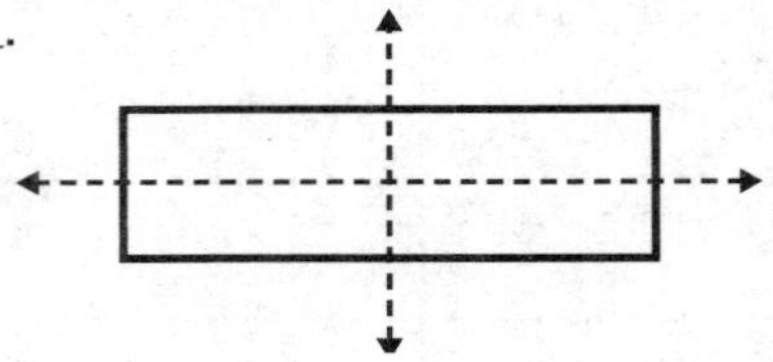

Figures can have ______ lines of symmetry.

Words I need help with:

My Math Examples:

NAME ______________________________ DATE ______________

Lesson 11 Problem-Solving Investigation

STRATEGY: Make a Model

Make a model to solve each problem.

1. **Mary Anne** is making a pattern with quadrilaterals.
   She put **squares** in the **first** row, **parallelograms** in the **second** row, and **trapezoids** in the **third** row.
   She **repeats** this **pattern four** times.
   Which **quadrilateral** does she use in the **tenth** row?

Understand	**Solve**
Plan I will draw the patterns to make a __________.	**Check**

2. Draw **two lines** on the square so that **three right triangles** are formed.

Understand	**Solve**
I know: A right triangle has ______ right angle. A square has _______ right angles. 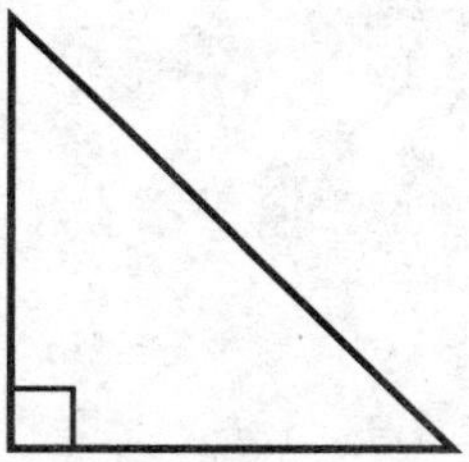	
Plan	**Check**

What are VKVs® and How Do I Create Them?

Dinah Zike's Visual Kinesthetic Vocabulary®

Visual Kinethestic Vocabulary Cards® are flashcards that animate words by focusing on their structure, use, and meaning. The VKVs in this book are used to show cognates, or words that are similar in Spanish and English.

Step 1

Go to the back of your book to find the VKVs for the chapter vocabulary you are currently studying. Follow the cutting and folding instructions at the top of the page. The vocabulary word on the BLUE background is written in English. The Spanish word is on the ORANGE background.

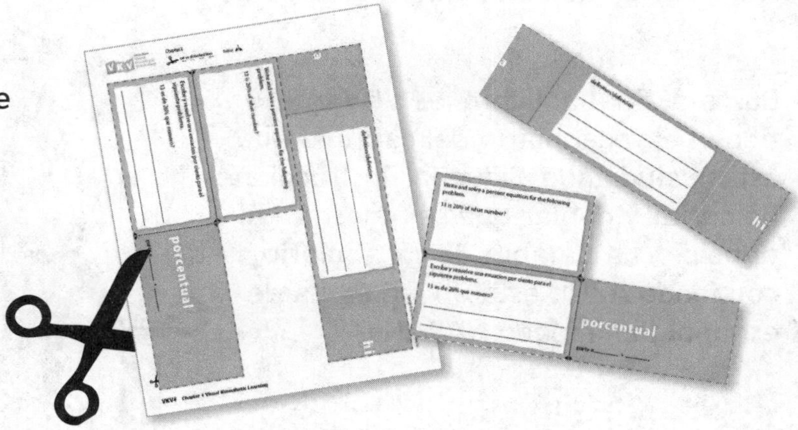

Step 2

There are exercises for you to complete on the VKVs. When you understand the concept, you can complete each exercise. All exercises are written in English and Spanish. You only need to give the answer once.

Step 3

Individualize your VKV by writing notes, sketching diagrams, recording examples, and forming plurals.

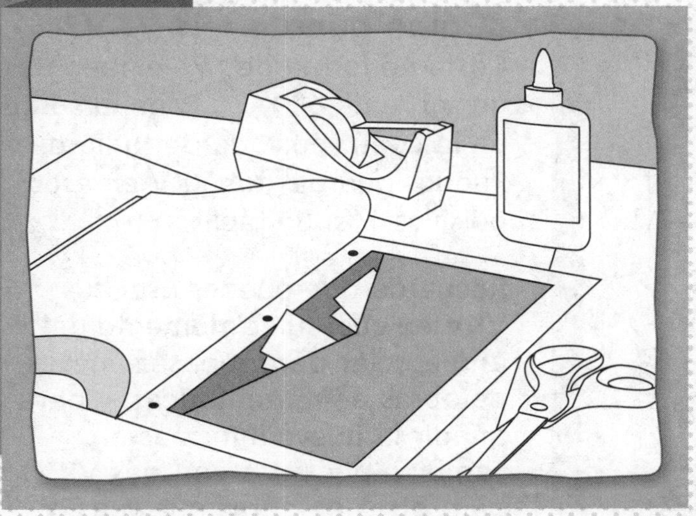

How Do I Store My VKVs?

Take a 6" x 9" envelope and cut away a V on one side only. Glue the envelope into the back cover of your book. Your VKVs can be stored in this pocket!

Remember you can use your VKVs ANY time in the school year to review new words in math, and add new information you learn. Why not create your own VKVs for other words you see and share them with others!

¿Qué son las VKV y cómo se crean?

Dinah Zike's Visual Kinesthetic Vocabulary®

Las tarjetas de vocabulario visual y cinético (VKV) contienen palabras con animación que está basada en la estructura, uso y significado de las palabras. Las tarjetas de este libro sirven para mostrar cognados, que son palabras similares en español y en inglés.

Paso 1

Busca al final del libro las VKV que tienen el vocabulario del capítulo que estás estudiando. Sigue las instrucciones de cortar y doblar que se muestran al principio. La palabra de vocabulario con fondo AZUL está en inglés. La de español tiene fondo NARANJA.

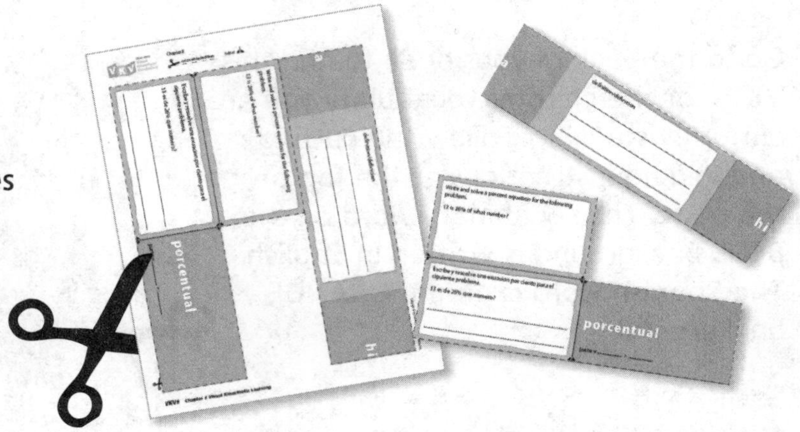

Paso 2

Hay ejercicios para que completes con las VKV. Cuando entiendas el concepto, puedes completar cada ejercicio. Todos los ejercicios están escritos en inglés y español. Solo tienes que dar la respuesta una vez.

Paso 3

Da tu toque personal a las VKV escribiendo notas, haciendo diagramas, grabando ejemplos y formando plurales.

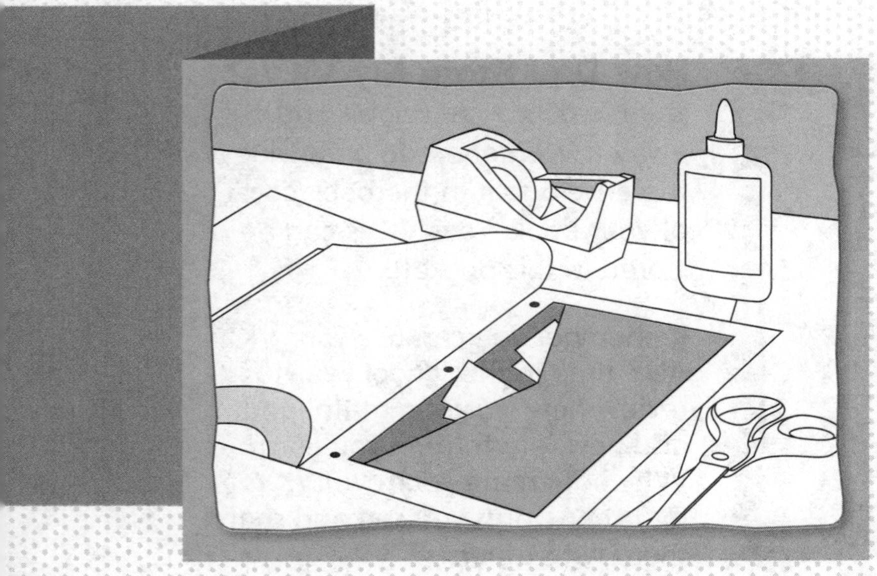

¿Cómo guardo mis VKV?

Corta en forma de "V" el lado de un sobre de 6" X 9". Pega el sobre en la contraportada de tu libro. Puedes guardar tus VKV en esos bolsillos. ¡Así de fácil!

Recuerda que puedes usar tus VKV en cualquier momento del año escolar para repasar nuevas palabras de matemáticas, y para añadir la nueva información. También puedes crear más VKV para otras palabras que veas, y poder compartirlas con los demás.

cut on all dashed lines

fold on all solid lines

digit

A digit is (Un dígito es) ____.

place value

A digit in each place represents ________ times what it would represent in the place to its right. (Un dígito en cada posición representa ________ veces lo que representaría en la posición de su derecha.)

What is the value of 7 in the number below? (¿Cuál es el valor de 7 en el número de abajo?)

3,758,241

____

period

Use commas to separate the periods in the number below. (Usa comas para separar los períodos en el número de abajo.)

3 2 5 4 8 2 2

cut on all dashed lines

fold on all solid lines

Write the highlighted digit's place and value. (Escribe el valor y la posición del dígito destacado.)

10,312 ______________________

____

154,065 ______________________

____

892 ______________________

____

9,473 ______________________

____

o

posicional

Thousands Period			Ones Period		
hundreds	tens	ones	hundreds	tens	ones
4	6	2	1	5	3

The 2 is in the ______ place. (El 2 está en la posición de ______.)

The 4 has a value of 4 × (El 4 tiene un valor de 4 ×) ______.

valor

o

Write the number below in standard form. Then circle the millions period and underline the thousands period. (Escribe el número de abajo en forma estándar. Luego, encierra en un círculo el período de los millones y subraya el período de los millares.)

Twenty-two million, five hundred sixty-three thousand

____

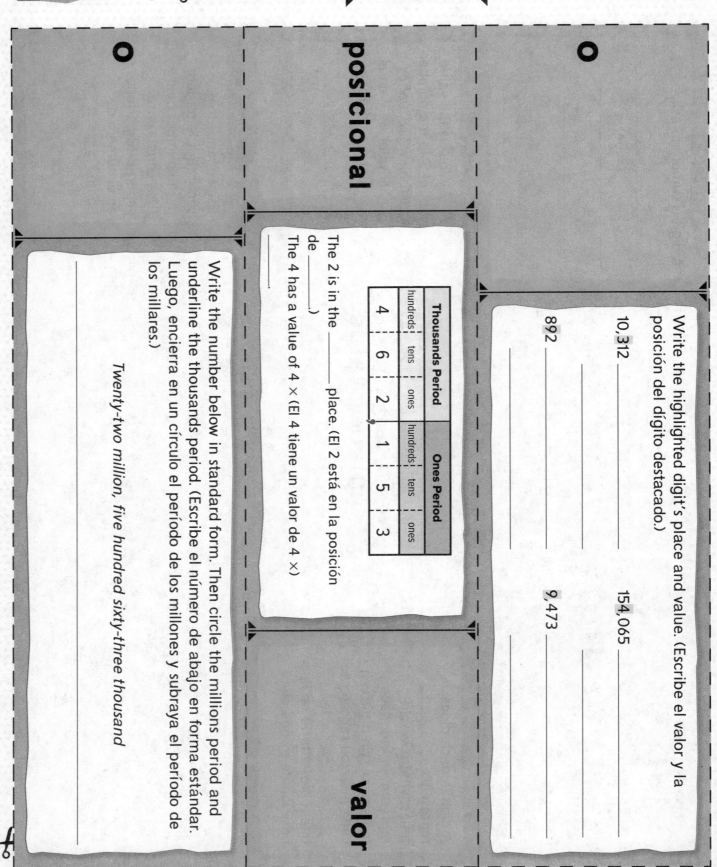

cut on all dashed lines

fold on all solid lines

Place a comma correctly into the number below. (Pon una coma correctamente en el número de abajo.)

2 5 4 6 5 8

standard form

Which shows standard form? (¿Cuál muestra la forma estándar?)

1. 5,000 + 200 + 10 + 6
2. *five thousand, two hundred sixteen*
3. 5,216

On a number line, numbers to the left are _______ than numbers to the right. (En una recta numérica, los números a la izquierda son _______ que los números a la derecha.)

number line

On a number line, numbers to the right are _______ than numbers to the left. (En una recta numérica, los números a la derecha son _______ que los números a la izquierda.)

Rounding is one way to _______ an answer. (Redondear es una manera de _______ una respuesta.)

round

Describe how to round a number. (Describe cómo redondear un número.) _______

____

____

____

estándar | **forma**

Write each number in standard form. (Escribe los números en forma estándar.)

50,000 + 4,000 + 200 + 60 + 1 ____________

three million, five hundred sixty-two thousand, seventy one ____________

600,000 + 30,000 + 500 + 7 ____________

numérica | **linea (recta)**

Use the number lines to compare. Use >, <, or =.
(Usa las rectas numéricas para comparar. Usa >, < o =.)

48,254 ◯ 44,365

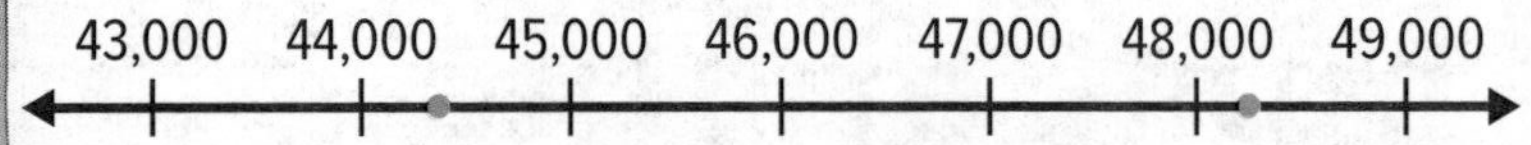

31,622 ◯ 31,854

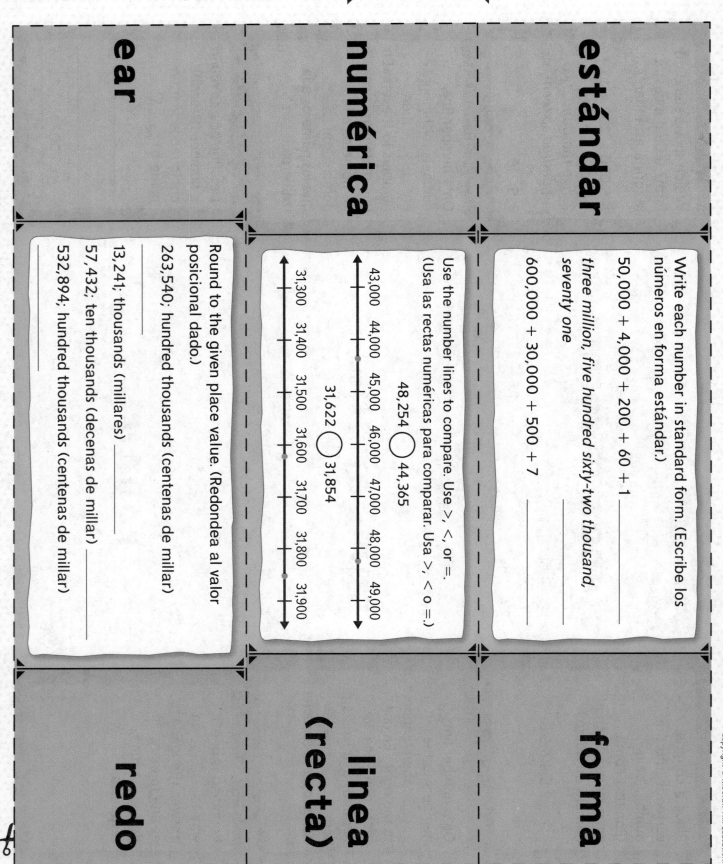

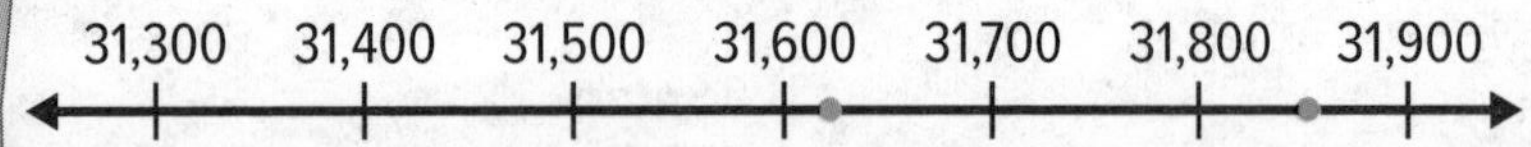

ear | **redo**

Round to the given place value. (Redondea al valor posicional dado.)

263,540; hundred thousands (centenas de millar) ____________

13,241; thousands (millares) ____________

57,432; ten thousands (decenas de millar) ____________

532,894; hundred thousands (centenas de millar) ____________

cut on all dashed lines

fold on all solid lines

Which example shows the Associative Property of Addition? Circle the answer. (¿Qué ejemplo muestra la propiedad asociativa de la suma? Encierra en un círculo la respuesta.)

1. 3 + (7 + 5) = (3 + 7) + 5

2. 6 + 4 = 10 and (y) 4 + 6 = 10

3. 5 + 0 = 5

Associative Property of Addition

conmutativa de la suma

Which example shows the Commutative Property of Addition? Circle the answer. (¿Qué ejemplo muestra la propiedad conmutativa de la suma? Encierra en un círculo la respuesta.)

1. $6 + (3 + 2) = (6 + 3) + 2$
2. $3 + 9 = 12$ and (y) $9 + 3 = 12$
3. $7 + 0 = 7$

propiedad asociativa de la suma

Commutative

cut on all dashed lines

fold on all solid lines

cut on all dashed lines

fold on all solid lines

Regroup 14 tens. (Reagrupa 14 decenas.)

_______ hundred +
_______ tens

regroup

Regroup 12 hundreds. (Reagrupa 12 centenas.)

_______ thousand +
_______ hundreds

Circle the subtrahend in the equation below. (Encierra en un círculo el sustraendo en la ecuación de abajo.)

$$\begin{array}{r} 5{,}326 \\ -\ 2{,}114 \\ \hline 3{,}212 \end{array}$$

subtrahend

The subtrahend is (El sustraendo es)

____
____
_______________.

An equation is a type of (Una ecuación es un tipo de) _______

____
_______________.

equation

Is *255* + *415* an equation? Why or why not? (¿Es *255* + *415* una ecuación? ¿Por qué?)

____
____

cut on all dashed lines

fold on all solid lines

ar | **reagr**

Use regrouping to add. (Usa el reagrupamiento para sumar.)

1,285	5,321	12,542
+ 184	+ 861	+19,375

o | **sustra**

At the beginning of May, Luke had $2,345. At the end of May, he had $1,574. How much did Luke spend? Write a subtraction problem to find the answer. Circle the subtrahend. (*A comienzos de mayo, Luke tenía $2,345. A fines de mayo, tenía $1,574. ¿Cuánto gastó Luke?* Escribe un problema de resta para hallar la respuesta. Encierra en un círculo el sustraendo.)

ción | **ec**

12 birds sat in a tree. 7 birds flew away. Then 8 more birds landed in the tree. How many birds are in the tree now? Write an equation with a variable to solve. (*12 aves estaban en un árbol. 7 se fueron volando. Luego, 8 aves más llegaron al árbol. ¿Cuántas aves hay ahora en el árbol?* Escribe una ecuación con una variable para resolver.)

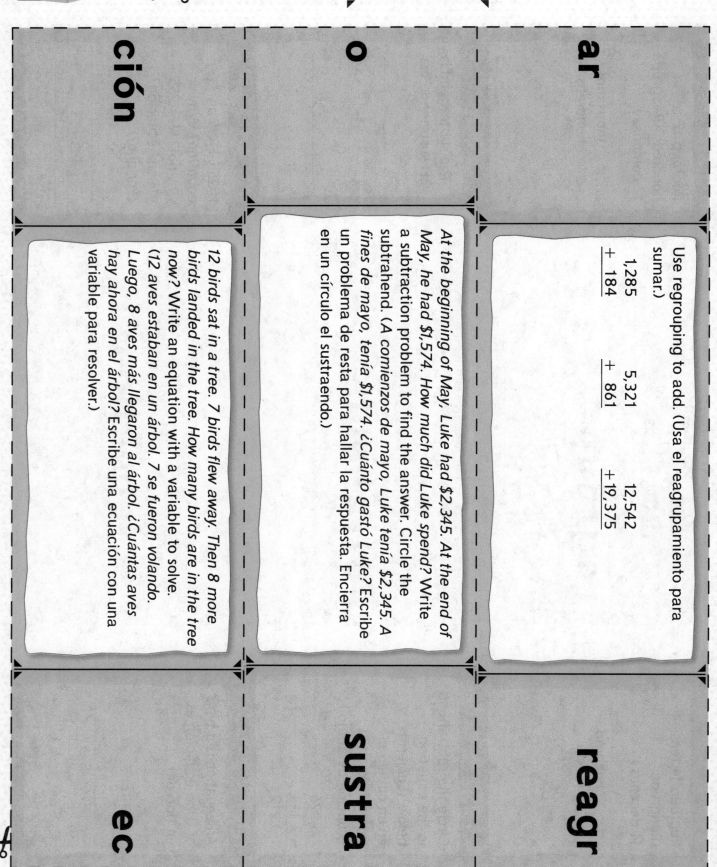

cut on all dashed lines

fold on all solid lines

Write a fact family for 7, 8, 56. Circle the quotients. (Escribe una familia de operaciones para 7, 8 y 56. Encierra en un círculo los cocientes.)

______ ______

______ ______

quotient

Write a fact family for 10, 12, 120. Circle the quotients. (Escribe una familia de operaciones para 10, 12 y 120. Encierra en un círculo los cocientes.)

______ ______

______ ______

factors

Circle the factors in each number sentence. (Encierra en un círculo los factores en cada oración numérica.)

$9 \times 5 = 45$

$12 \times 12 = 144$

$6 \times 3 = 18$

multiplication

Label each part of the multiplication sentence. (Rotula cada parte de la multiplicación.)

$6 \times 9 = 54$

______ ______ ______

cut on all dashed lines

fold on all solid lines

e

When you write a multiplication sentence related to a division sentence, will the quotient become a factor or product? (Cuando escribes una multiplicación relacionada a una división, ¿se convierte el cociente en factor o en producto?)

____

____

coc

es

Find each unknown to complete the fact family. Circle the factors. (Halla las incógnitas para completar la familia de operaciones. Encierra en un círculo los factores.)

$12 \times$ ____ $= 72$	____ $\times 4 = 28$
$6 \times 12 =$ ____	____ $\times 7 = 28$
____ $\div 6 = 12$	____ $\div 7 = 4$
$72 \div$ ____ $= 6$	$28 \div 4 =$ ____

ción

Hector dealt 7 playing cards to 4 friends. How many cards did he deal altogether? Write a multiplication sentence to solve. (*Héctor repartió 7 tarjetas de un juego a 4 amigos. ¿Cuántas tarjetas repartió en total?* Escribe una multiplicación para resolver.)

____

____

____

cut on all dashed lines

fold on all solid lines

propiedad conmutativa

Associative Property of Multiplication

Use the Associative Property of Multiplication to show another way to solve. (Usa la propiedad asociativa de la multiplicación para mostrar otra forma de resolver.)

$2 \times 5 \times 7 = 2 \times (5 \times 7)$	$2 \times 5 \times 7 =$ ________
$= 2 \times 35$	$=$ ________
$= 70$	$=$ ________

cut on all dashed lines

fold on all solid lines

Which example shows the Commutative Property of Multiplication? Circle the answer. (¿Qué ejemplos muestran la propiedad conmutativa de la multiplicación? Encierra en un círculo la respuesta.)

1. $12 \times 5 = 5 \times 12$

2. $18 \times 1 = 18$

3. $9 \times 0 = 0$

propiedad asociativa de la multiplicación

Commutative Property

cut on all dashed lines

fold on all solid lines

Compatible numbers are (Los números compatibles son)

____

______________.

compatible numbers

What is the nonmath meaning of *partial*? (¿Cuál es el significado no matemático de *parcial*?)

____

partial quotients

List 3 partial quotients to help solve $848 \div 4$. (Menciona 3 cocientes parciales que te ayuden a resolver $848 \div 4$.)

_____ _____ _____

cut on all dashed lines

fold on all solid lines

s

Use compatible numbers and mental math to estimate. Then check your estimates. (Usa números compatibles y cálculo mental para estimar. Luego, comprueba tus estimaciones.)

1,197 ÷ 3 is about (es aproximadamente) _____. 1,197 ÷ 3 = _____

88 × 5 is about (es aproximadamente) _____. 88 × 5 = _____

523 × 7 is about (es aproximadamente) _____. 523 × 7 = _____

números

parciales

Divide using partial quotients.
(Divide usando cocientes parciales.)

612 ÷ 3 355 ÷ 5

cocientes

cut on all dashed lines

fold on all solid lines

numeric pattern

What rule does the pattern follow? (¿Qué regla sigue el patrón?)

12, 17, 22, 27, 32

____

order of operations

Find the value of each expression. (Halla el valor de cada expresión.)

$(3 \times 5) - (6 \div 2) =$ ______

$10 + (25 \div 5) =$ ______

cut on all dashed lines

fold on all solid lines

patrón

Find the unknown in each numeric pattern.
(Halla la incógnita en cada patrón numérico.)

8, 12, 13, 17, 18, 22, 23, ____

55, 75, 65, ____, 75, 95, 85, 105

8, ____, 32, 64, 128

o

orden de las

Write 1–3 on the lines to show the correct order of operations. (Escribe 1, 2 y 3 en la líneas para mostrar el orden de las operaciones correcto.)

____ Add and subtract in order from left to right. (Suma y resta en orden de izquierda a derecha.)

____ Perform operations in parentheses. (Haz las operaciones que están entre paréntesis.)

____ Multiply and divide in order from left to right. (Multiplica y divide en orden de izquierda a derecha.)

ciones

cut on all dashed lines

fold on all solid lines

Name two nonmath examples of pairs. (Nombra dos ejemplos no matemáticos de pares.)

____

____

factor pairs

List two factor pairs for 12. (Menciona dos pares de factores para 12.)

_____ and (y) _____

_____ and (y) _____

Define *prime number*. (Define *número primo*.)

____

____

____

____

prime number

Circle each prime number. (Encierra en un círculo cada número primo.)

3	21	2	35
17	77	14	39

fraction

Write a fraction with a denominator of 9 and a numerator of 5. (Escribe una fracción cuyo denominador sea 9 y el numerador sea 5.)

cut on all dashed lines
fold on all solid lines

factores | **pares de**

Find the factor pairs for each number.
(Halla los pares de factores para cada número.)

54 ______ × ______ ______ × ______

______ × ______ ______ × ______

26 ______ × ______ ______ × ______

primo | **número**

A prime number has ______ factors.
(Un número primo tiene ______ factores.)

A composite number has ______ factors.
(Un número compuesto tiene ______ factores.)

ción

Circle the fractions that are equivalent in each set. (Encierra en un círculo las fracciones que sean equivalentes en cada conjunto.)

1. $\frac{5}{6}$ $\frac{15}{18}$ $\frac{10}{18}$

2. $\frac{10}{40}$ $\frac{12}{36}$ $\frac{1}{4}$

3. $\frac{1}{2}$ $\frac{14}{28}$ $\frac{27}{56}$

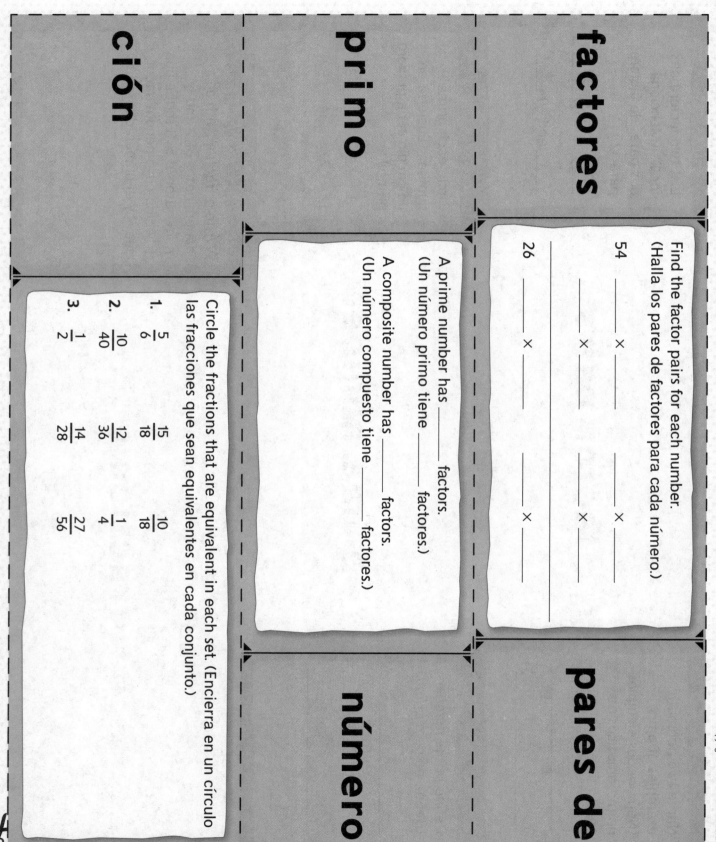

cut on all dashed lines

fold on all solid lines

least common multiple

Describe why you might need to find the least common multiple of two numbers. (Describe por qué podrías necesitar hallar el mínimo común múltiplo de dos números.)

____

____

____

Write the greatest common factor (GCF) of the numerator and denominator. Then write the fraction in simplest form. (Escribe el máximo común divisor (M.C.D.) del numerador y el denominador. Luego, escribe la fracción en su mínima expresión.)

$\frac{15}{18}$

GCF: ______

simplest form: ______

$\frac{21}{28}$

GCF: ______

simplest form: ______

máximo común divisor

cut on all dashed lines

fold on all solid lines

greatest common factor

List the factors for each number. Circle the greatest common factor. (Menciona los factores de cada número. Encierra en un círculo el máximo común divisor.)

12: ___, ___, ___, ___, ___, ___

18: ___, ___, ___, ___, ___, ___

Write the least common multiple for each pair of numbers. (Escribe el mínimo común múltiplo de cada par de números.)

5 and (y) 7 ___   2 and (y) 8 ___

3 and (y) 8 ___   2 and (y) 5 ___

Circle the greatest fraction. (Encierra en un círculo la fracción más grande.)

$\frac{1}{2}$ $\frac{12}{25}$ $\frac{3}{5}$

mínimo común múltiplo

cut on all dashed lines

fold on all solid lines

A unit fraction has a numerator of (Una fracción unitaria tiene un numerador de)

________.

unit fraction

Add the unit fractions. (Suma las fracciones unitarias.)

$\frac{1}{5} + \frac{1}{5} + \frac{1}{5} =$ ____

simplify

When you simplify, you write a fraction in its ________ form. (Al simplificar, una fracción se escribe en su ________ expresión.)

Write $6\frac{1}{3}$ as an improper fraction. (Escribe $6\frac{1}{3}$ como fracción impropia.)

improper fraction

Write $\frac{5}{2}$ as a mixed number. (Escribe $\frac{5}{2}$ como número mixto.)

cut on all dashed lines

fold on all solid lines

unitaria

Use the model to find $\frac{1}{8} + \frac{3}{8}$. (Usa el modelo para hallar $\frac{1}{8} + \frac{3}{8}$.)

How many unit fractions did you shade? (Spanish translation) _____

fracción

icar

Simplify each fraction. (Simplifica las fracciones.)

$\frac{12}{24}$ = _____

$\frac{4}{16}$ = _____

$\frac{9}{81}$ = _____

$\frac{10}{60}$ = _____

$\frac{24}{36}$ = _____

$\frac{12}{21}$ = _____

impropia

Describe how to write a mixed number as an equivalent improper fraction. (Describe cómo escribir un número mixto como una fracción impropia equivalente.)

_____

_____

_____

fracción

cut on all dashed lines

fold on all solid lines

convert

Use *convert* in a nonmath sentence. (Usa *convertir* en una oración no matemática.)

mile

List two things you would measure in miles. (Menciona dos cosas que medirías en millas.)

capacity

Capacity is (La capacidad es) ______

ir

How would you convert 12 yards into feet? (¿Cómo convertirías 12 yardas a pies?)

____

How would you convert 24 inches into feet? (¿Cómo convertirías 24 pulgadas a pies?)

____

la

Mr. Diaz rides his bike 2 miles to work each morning. He rides home again in the afternoon. How far in feet does he ride each day? (El Sr. Díaz anda en bicicleta 2 millas hasta el trabajo todas las mañanas. Vuelve a su casa en bicicleta por la tarde. ¿Cuánto anda en bicicleta cada día?)

1 mile (mi) = 5,280 feet (ft)

____

dad

Write the units of capacity in order from least to greatest. (Escribe las unidades de capacidad en orden de menor a mayor.)

quart	cup	ounce	gallon	pint
cuarto	taza	onza	galón	pinta

____

cut on all dashed lines

fold on all solid lines

There are ________ pints in 1 quart.

(Hay ________ pintas en 1 cuarto.)

quart

There are ________ quarts in 1 gallon.

(Hay ________ cuartos en 1 galón.)

ton

There are ________ pounds in 2 tons.

(Hay ________ libras en 2 toneladas.)

seconds

Write an expression that shows the number of seconds in a day. (Escribe una expresión que muestre el número de segundos que hay en un día.)

____

cut on all dashed lines

fold on all solid lines

o

Circle the most reasonable estimate for capacity. (Encierra en un círculo la estimación más razonable de la capacidad.)

1 quart (1 cuarto   8 quarts 8 cuartos   25 quarts 25 cuartos   50 quarts 50 cuartos)

c

elada

Write a real-world problem involving tons. (Escribe un problema del mundo real en el que haya toneladas.)

____

____

____

____

gundos

Convert each time measurement into seconds. (Convierte las mediciones del tiempo en segundos.)

2 hours and 14 minutes = ________ seconds
(2 horas y 14 minutos = ________ segundos)
154 minutes = ________ seconds
(154 minutos = ________ segundos)
1 hour and 55 minutes = ________ seconds
(1 horas y 55 minutos = ________ segundos)

cut on all dashed lines

fold on all solid lines

Circle the word part in *centimeter* that means "hundred." (Encierra en un círculo la parte de la palabra *centímetro* que significa "cien".)

centimeter

_____ mm = 1cm

1m = _____ cm

A milliliter is a _______ unit of capacity.

(Un mililitro es una unidad _______ de capacidad.)

milliliter

_____ mL = 1L

Circle the word part in *kilogram* that means "thousand." (Encierra en un círculo la parte de la palabra *kilogramo* que significa "mil".)

kilogram

How many grams are there in 1 kilogram? (¿Cuántos gramos hay en 1 kilogramo?)

____

ro

Circle the most reasonable estimate for length. (Encierra en un círculo la estimación más razonable de la longitud.)

paperclip (clip para papel)	3 cm	30 cm	300 cm
book (libro)	20 cm	200 cm	2,000 cm
tree (árbol)	10 cm	100 cm	1,000 cm

centí

ro

Circle the better unit to measure each capacity. (Encierra en un círculo la mejor unidad para medir cada capacidad.)

juice box (caja de jugo)	milliliter (mililitro)	liter (litro)
fish tank (pecera)	milliliter (mililitro)	liter (litro)
ink bottle (tintero)	milliliter (mililitro)	liter (litro)
bucket (cubeta)	milliliter (mililitro)	liter (litro)

mi

o

List 4 objects that weigh more than 1 kilogram. (Menciona 4 objetos que pesen más de 1 kilogramo.)

__________________ __________________

__________________ __________________

kiló

cut on all dashed lines

fold on all solid lines

Parallel lines are always (Las líneas paralelas siempre son) ______________

parallel

An obtuse angle measures (Un ángulo obtuso mide) ______ than 180°. (de 180°.)

obtuse angle

An obtuse angle measures (Un ángulo obtuso mide) ______ than 90°. (de 90°.)

What symbol indicates a right angle? (¿Qué símbolo indica un ángulo recto?) ______________

right angle

A right angle measures (Un ángulo recto mide) ______ degrees. (grados.)

cut on all dashed lines

fold on all solid lines

ela

Draw an example of the figure.
(Dibuja un ejemplo de la figura.)

$\overline{AB} \parallel \overline{DE}$

obtuso

Draw two different examples of obtuse angles.
(Dibuja dos ejemplos diferentes de ángulos obtusos.)

ángulo

recto

Draw a four-sided shape with four right angles.
(Dibuja una figura de cuatro lados con cuatro ángulos rectos.)

ángulo

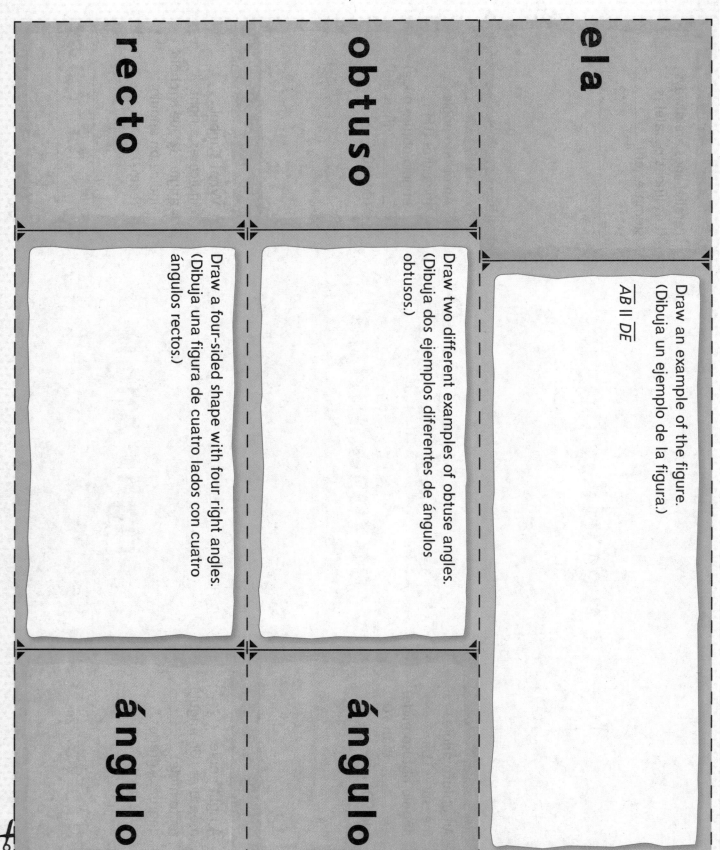

cut on all dashed lines

fold on all solid lines

An acute triangle has ______ acute angle(s). Draw an example of an acute triangle.

(Un triángulo acutángulo tiene ______ ángulo(s) agudo(s). Dibuja un ejemplo de un triángulo acutángulo.)

acute triangle

rectángulo

cut on all dashed lines

fold on all solid lines

triángulo agudo

right

A right triangle has ______ right angle(s). Draw an example of a right triangle. Be sure to indicate the right angle. (Un triángulo rectángulo tiene ______ ángulo(s) recto(s). Dibuja un ejemplo de un triángulo rectángulo. Asegúrate de indicar el ángulo recto.)